"I PROMISE YOU NO SUGARCOATING. THERE ARE SERIOUS FACTORS IN AVIATION THAT PRODUCE SERIOUS— AND UNNECESSARY—RISKS. THEY ARE DOCUMENTED HERE. . . . BUT I ALSO PROMISE THAT YOU WILL FEEL BETTER ABOUT FLYING AFTER READING THIS BOOK. . . . AND SOME OF YOU WHO ALREADY LIKE IT MAY COME TO LOVE IT." —*F. Lee Bailey*

When F. Lee Bailey takes on a job, he gives his all—as those whom he has defended in court, and those who have read his riveting bestsellers, can testify.

Now F. Lee Bailey has taken on one of his greatest challenges—to investigate, evaluate, and make recommendations for safety aboard our modern jets.

He has given it all of his investigative energy, his untiring pursuit of the truth. And he has given us all a book that we can read for pleasure and for knowledge—and should read before we board a plane again!

"IMPORTANT!"—HOUSTON CHRONICLE

SIGNET and MENTOR Books of Special Interest

☐ **THE DEFENSE NEVER RESTS by F. Lee Bailey with Harvey Aronson.** The million-copy bestseller that presents the most sensational cases of the decade! "Splendid, blunt, fascinating . . . highly recommended!"—Newsweek (#E8317—$2.25)

☐ **FOR THE DEFENSE by F. Lee Bailey with John Greenya.** Unforgettable trials are brought to life as F. Lee Bailey takes you into a brilliant lawyer's mind—and with him you face the greatest courtroom challenges of our time. "A great book by a great trial lawyer!"—Vincent T. Bugliosi, co-author of Helter Skelter (#J7022—$1.95)

☐ **BRING ME A UNICORN: The Diaries and Letters of Anne Morrow Lindbergh (1922-1928) by Anne Morrow Lindbergh.** Imagine being loved by the most worshipped hero on earth. This story of Charles Lindbergh and Anne Morrow Lindbergh is the chronicle of just such a love. "Extraordinary . . . brings to intense life every moment as she lived it."—The New York Times Book Review (#E8447—$1.75)

☐ **HOUR OF GOLD, HOUR OF LEAD by Anne Morrow Lindbergh.** The Lindberghs were the golden couple in a fairy-tale romance. And when their first child was born, the world rejoiced. Eighteen months later, tragedy struck. . . . "A totally expressive, often unbearable record of an extreme personal anguish that followed the greatest possible happiness. Mrs. Lindbergh has a great gift for communicating directly her joy and pain."—The New York Times Book Review (#E5825—$1.75)

☐ **THE NEW MUCKRAKERS by Leonard Downie, Jr.** An inside look at the new American heroes—the star investigative reporters . . . "The best book on the subject!"—The New York Times Book Review (#ME1628—$2.50)

If you wish to order these titles,

please see the coupon

in the back of this book.

CLEARED FOR
THE APPROACH

CLEARED FOR THE APPROACH:
F. Lee Bailey In Defense of Flying

F. LEE BAILEY with John Greenya

UPDATED AND WITH A NEW EPILOGUE

A SIGNET BOOK
NEW AMERICAN LIBRARY
TIMES MIRROR

NAL BOOKS ARE ALSO AVAILABLE AT DISCOUNTS IN BULK
QUANTITY FOR INDUSTRIAL OR SALES-PROMOTIONAL USE.
FOR DETAILS, WRITE TO PREMIUM MARKETING DIVISION,
NEW AMERICAN LIBRARY, INC., 1301 AVENUE OF THE
AMERICAS, NEW YORK, NEW YORK 10019.

COPYRIGHT © 1977, 1978 BY F. LEE BAILEY

All rights reserved. No part of this book may be reproduced
in any form or by any means, except for the inclusion of brief
quotations in a review, without permission in writing from the
publisher. For information address The New American Library, Inc.

Library of Congress Catalog Card Number: 77-23151

A hardcover edition was published by Prentice-Hall, Inc.

 SIGNET TRADEMARK REG. U.S. PAT. OFF. AND FOREIGN COUNTRIES
REGISTERED TRADEMARK—MARCA REGISTRADA
HECHO EN CHICAGO, U.S.A.

SIGNET, SIGNET CLASSICS, MENTOR, PLUME AND MERIDIAN BOOKS
are published by The New American Library, Inc.,
1301 Avenue of the Americas, New York, New York 10019

FIRST SIGNET PRINTING, OCTOBER, 1978

1 2 3 4 5 6 7 8 9

PRINTED IN THE UNITED STATES OF AMERICA

Introduction

Opening Statement

An introduction to a book is much like an opening statement to a jury: You make promises. In court you tell the jurors what your side of the story will be, and if the evidence unfolds as you want it to, you normally prevail. Things can go wrong, however; witnesses have an unpleasant habit of dying at the wrong time, or being struck by a sudden memory loss, and judges have been known to turn a deaf ear now and then. Thus you have to be rather careful what you tell a jury at the very beginning, because you may not be in total control of the events that will follow.

The writing of books (praise the Lord) is different, in that the writer is the one who decides what "evidence" goes in and what does not. And if he or she can't back up a claim made in the introduction, he or she had better not make it. Thus anyone who includes an introduction is confident that the chapters that follow will logically keep the early promises.

It is not quite *that* simple, however. Just as a juror can be moved by a subconscious bias, a reader can be swayed by an illogical, wholly emotional reaction to a given subject. One of the greatest of these swaying factors is fear.

Good old clean irrational fear—it has kept all of us at one time or another from doing something we might otherwise have enjoyed. Some people, although they have all the requisite mental and physical skills, are afraid to learn to drive a car. Others would not get on a horse for all the money in Howard Hughes's will. For some people a simple boat ride is an invitation to disaster. And then we come to flying.

Twenty years ago, the majority of Americans had not yet

[viii] *CLEARED FOR THE APPROACH*

had their first airplane ride. Today, the person who has never flown is considered to be unusual. Generally speaking, everyone flies nowadays. *But* everyone does not like it. For a variety of reasons, otherwise intelligent people with the normal amount of fortitude are damn nervous when they get up in an airplane. You can talk until you're blue in the face explaining the principles of flight, and all you get in return is a glazed look that, translated into words, means, "But what is this big heavy thing doing up in the air?"

I sympathize with such people, if for no other reason than that I have met so many of them. But I cannot and do not agree with them. Flying is so much safer than driving that it seems silly to debate the point, or even to bring it up. But that fear exists, and any book that purports to deal with flying and does not recognize that fact would misserve its readers.

So, I make the first of my promises. I will explain to you why you should not be afraid of flying per se.

I also promise that you will get no sugarcoating. There are factors involved in the totality of aviation that produce serious —and unnecessary—risks; they are documented here. However, I promise not to frighten you unduly.

A grab bag promise: You will feel better about flying after you read this book. Some of you may even come to like it, and some of you who already like it may come to love it.

A word, then, about credentials—I have been flying airplanes far longer than I have been trying lawsuits. I became a pilot at the age of twenty, thanks to the United States Navy and, later, the Marines. I have flown my own airplanes, from Cessnas to Lear Jets. I am a licensed helicopter pilot. I once owned and operated a small airport, and I have taught others to fly. I presently operate an air charter service, and I now own a company that makes helicopters.

If the law has been my life, then flying has been my avocation. As a lawyer, I have a firm that tries air crash cases; as an owner of aircraft and a pilot, I have encountered just about every legal situation in the regulations; as a manufacturer, I have come to know federal requirements by heart. With the exception of airline crew members, I don't think there are very many Americans who have logged more air miles than I have. I am known as a defense lawyer, but I would be hard put to say which has consumed more of my time in the past two decades, being a lawyer or being a pilot.

Finally, a word about the way this book is arranged. This is not a book for other pilots or people in the aviation field alone; in fact, it is primarily a book for the *consumer* of America's aviation product—the air traveler. Thus the order is the reverse of what I would use in different circumstances. The first and more lengthy section deals with the fear of flying. This section explores "what worries people," and it looks at the components of the system. It covers both cause for concern and cause for calm.

The second section deals with the love of flying—why I love it, why others do, and why you might just come to feel that way too.

The book begins on a grisly note, with the last moments of a doomed airplane. I chose to begin that way for a very simple reason: that crash, like all but the most bizarre of accidents, need not have happened. Once you have seen what went wrong, you can understand why it should not have, and why it so seldom does. I'll take you through a schematic that includes the people who make the planes, those who fly them, those who regulate them, and those who make sure they get where they are going. When you're finished, you will understand why so many people in this country feel safer in the air.

F. LEE BAILEY

Marshfield, Massachusettts

Contents

INTRODUCTION vii

Part I
Fear of Flying

1. "Sound of Impact" 3
2. The Industry 18
3. The Government 33
4. The Human Factor 61
5. Natural Factors 116
6. "After the Fall":
   The Trying of An Air Crash Case 148

Part II
Love of Flying

1. Love of Flying 163
2. The World of General Aviation 185

RECOMMENDATIONS 199

EPILOGUE 213

1104:16—TWA five fourteen, you're cleared for a VOR DME approach to runway one two.

1104:21—Cleared for a DME, uh VOR DME approach runway one two, TWA five fourteen, roger.

CLEARED FOR
THE APPROACH

PART I
FEAR OF FLYING

1

"Sound of Impact"

When the weather is bad, as it was on the morning of December 1, 1974, the cockpit of a Boeing 727 is no place for casual conversation. The wind howls and the rain slams against the glass. Crew members have to raise their voices a bit, which is what the captain did, at ten minutes to eleven, Eastern Daylight Time, when he said to his flight engineer, "Did you tell the people yet?"

The engineer knew that the eighty-five passengers, "the people," weren't going to be happy when they heard his announcement. Their tickets said that they would land at Washington's National Airport. The copilot knew it too, and he joked, "Ladies and Gentlemen, for those of you who live near Dulles, we got some good words for you. The good news is, uh, if you live near Dulles, we're gonna land there. The bad news is, if you live near National, you gonna be on at Dulles."

"Yeah," kidded the engineer, "you got a bus ride coming." Then he punched the button and told the people.

* * *

Throughout the Washington area that morning, sensible people were staying indoors. The rain was bad enough, but the strong winds were felling trees and taking power lines down. Pepco had emergency crews all over the District, and across the Potomac River at National Airport, the winds were sweeping the runways at 40 to 45 knots. Neither Captain Richard Brock, nor his first officer, Leonard Kresheck, who was flying the airplane that morning, had any desire to test their combined skill against crosswinds of such intensity. 514 would divert to Dulles.

Once the decision had been made, the captain relaxed enough to ask the flight engineer to hand him his breakfast tray, which had been sitting on the engineer's table for almost ten minutes. Captain Brock was not impressed with the food. "Aw, look at that sick-looking crap."

Tom Sefranek, the engineer, decided, "I think I'm going to have time to eat at Dulles."

By eleven o'clock, 514 had been in the air for only thirty-three minutes. The flight, which had originated in Indianapolis, Indiana, had made a scheduled intermediate stop at Columbus, Ohio, leaving there at 10:27, Eastern Daylight Time. But it had not been an easy flight. Already, 514 had encountered heavy winds, all but constant rain, and occasional snow.

The turbulence was officially "moderate," but it was more than enough to worry the passengers. A few minutes before eleven o'clock, the first officer had suggested to the captain, "I think the girls ought to plunk their asses down too." The captain agreed, and within seconds the seat belt and no-smoking signs were turned on. In the passenger sections there was no liquor to be seen, and all the seat back trays were up and fastened.

The captain activated the ignition, the anti-ice mechanism, and the logo light.

The first officer mentioned an earlier flight into Dulles: "You know they get this once in a while. We went in here at Dulles one time before. We diverted out of National, went in to Dulles—same thing—the wind was blowing like a —————, it just beat the ————— out of us going down there."

Captain Brock was already thinking ahead. "Well, if we can't make it for any reason, we'll just have to go to Kennedy." But he didn't expect anything unusual, ". . . unless it's for some dumb reason like another airplane on the runway or something."

* * *

Shortly after eleven the wind picked up again. Inside the cockpit, the crew could hear the sound of heavy rain mixed

FEAR OF FLYING [5]

with sleet pounding against the aluminum skin of the 727. The captain noticed, ". . . a lot of snow around here . . ."

Soon it got so bumpy that one of the crew actually laughed. The engineer said the wind was "zero eight zero at two five gusting three six."

Then, at exactly four minutes and nine seconds after eleven, 514 contacted approach control at Dulles.

Seven seconds later, the controller monitoring 514's approach to Dulles answered, "TWA five fourteen, you're cleared for a VOR DME approach to runway one two."

The engineer responded, "Cleared for a DME, uh, VOR DME approach runway one two, TWA five fourteen, roger."

With that, 514 was going to land on runway 12 at Dulles in approximately seven minutes, unless there was an emergency, or something unforeseen, some "dumb reason."

* * *

Among the passengers were several people who weren't accustomed to running behind schedule. One of them was Roscoe C. Cartwright, a handsome black man of late middle age, who was traveling with his wife; a Brigadier General in the United States Army, Cartwright knew that if it had been a military flight, it would not have been twenty-three minutes late in taking off from Indianapolis.

There were two FBI agents on board, and neither was fond of tardiness. But they, like the Cartwrights, knew that if the professionals up front felt it wasn't safe to land at National, then that was the correct decision. There were also several congressional aides on board, but they were more than familiar with unavoidable delays.

By eleven o'clock the passengers knew they were going to Dulles, and they knew that they were already almost a half hour late. What they didn't know was that there was some confusion up front.

First Officer Kresheck the copilot, was flying the plane (that was not at all unusual; it is the way copilots build up enough experience to become captains), which meant that he was actually handling the controls and scanning some of the instruments. At the same time, Captain Brock was monitoring all the instruments and watching out the window for visual checkpoints. Flight Engineer Sefranek was watching the gauges to make sure the engines were functioning properly, monitoring the rate of descent, and making sure there were no deviations from the approach pattern.

In addition to his usual duties, the captain was also reading the approach plate, the "map" that tells how to approach every large airport in the country. At four minutes and twenty-seven seconds after eleven, he said, "Eighteen hundred's the bottom."

"Start down," said Kresheck, indicating he was beginning an immediate descent to the initial approach altitude, the minimum safe altitude prior to landing.

No one said anything for thirty seconds, but when the force of the wind again became noticeable, the first officer said, "You can feel that wind down here now . . . not that you can do anything about it."

Captain Brock did not respond to the comment about the wind. He had just read something that bothered him. "You know," he said, "according to this dumb sheet, it says that 3400 to Round Hill is our minimum altitude."

Moments later, Sefranek said, "Where do you see that?"

"Well, here," said the captain, indicating the approach plate spread open in his lap, "Round Hill is eleven and a half DME."

Kresheck started to say something, but waited, perhaps deferring to Brock's nineteen years of service with TWA, as opposed to his own nine years. When Captain Brock spoke again, the usual authority was back in his voice: "When he [the air traffic controller] clears you, that means you can go to your initial approach altitude." That settled the matter.

At eight and a half minutes after the hour, the captain noticed Kresheck struggling with the controls. He said, "Hang in there, boy."

The engineer said, "We're getting seasick."

Eleven oh nine, minus three seconds: The altitude alert horn sounded in the cockpit.

Kresheck said, "Boy, it was . . . wanted to go right down through there, man. . . . Must have had a ─────── of a downdraft."

Seventeen seconds later the radio altimeter warning horn sounded, then stopped. The engineer said, "Boy!"

Six seconds later, Brock yelled, "Get some power on!" Again the radio altimeter warning horn sounded and stopped.

Then, at exactly nine minutes and twenty-two seconds after

eleven, TWA Flight 514 crashed into the side of a small mountain.

Everyone on board was killed instantly.

Less than a week later, an employee of the National Transportation Safety Board finished transcribing the tape from the cockpit voice recorder, a device used to record the last 30 minutes of cabin talk during all commercial flights. The final entry in the thirty-five page document read: "1109:22—Sound of Impact."

* * *

At the precise moment that the red, white, and silver airplane sheared off the treetops and crashed into a huge outcropping of rock on the west side of Mount Weather, Washington National Airport was slowly filling up with people.

The Thanksgiving weekend was over, and thousands of people were scheduled to arrive or depart, and thousands more were picking them up or seeing them off.

Quite often on holiday weekends, an airport is a pleasant if hectic place to be. The festive mood is contagious, and people seem almost to enjoy the bustling pace. But that wasn't the way it was at National Airport on the first Sunday of December 1974. The weather was getting worse, not better, and the travelers' moods reflected the changes. Each time new flight information blinked onto the closed television screens and showed yet another delay, more people grew restless.

Those who bothered to stare out the windows onto the field saw nothing but rain and the effects of a harsh, driving wind. No one used the pay telescopes on the open balcony. Occasionally, a ground crew member would appear walking with difficulty against the wind, one hand stretched out for balance, the other clutching his cap.

On the south end of the terminal in the new wing where TWA is located, people jammed the counters asking for information. By noon, the clerks were already looking harried, already beginning to snap at people, some of whom were by then two hours behind schedule.

Out at the gates, the waiting areas for passengers, the people milled around or sat with bored expressions. One large group was waiting for flight 427, the return leg of the incoming 514. Among this group was Indiana's junior senator,

[8] *CLEARED FOR THE APPROACH*

Birch Bayh. It was ironic that of all the United States senators who might have been waiting at that airport on that day, Birch Bayh should be the one. For it was Bayh (and his wife) who accompanied Teddy Kennedy on an ill-fated private flight in 1962 that crashed, killing the pilot and severely injuring Kennedy, whom Bayh helped to pull from the wreckage.

In addition to the lone politician there was a group of professional politician-watchers. Because the Republican Governors' Conference and the Democrats' "Mini-Convention" were both being held in the Midwest that week, an unusually large group of political reporters was waiting for several different TWA flights. Among them were two UPI men, the wife of an AP reporter, and a number of Washington bureau chiefs representing papers all over the country. Nationally syndicated columnist Rowland Evans was one of the group.

Some of the reporters talked about the recent odd behavior of Congressman Wilbur Mills—who at that very moment was making arrangements to charter a Lear Jet so that he could fly up to Boston that evening and appear on stage with his friend Fanne Foxe. (The next morning's Washington *Post* would feature two headline stories: the Mills' appearance and the crash of TWA 514.)

One reporter, Adam Clymer of the Baltimore *Sun*, sat quietly with a small group of nervous-looking little girls, aged eight to twelve, from St. Joseph's School for the Deaf in St. Louis. One of the girls was the daughter of a colleague, and Clymer had promised to keep an eye on the children. Occasionally one of the other reporters would walk over to talk with Clymer and smile at the girls, who were holding elaborately wrapped Christmas gifts purchased during their stay in Washington.

In the gate adjacent to where 514 had been scheduled to arrive, two UPI reporters, Arnie Sawsilak and Steve Gerstel, had been waiting for ninety minutes for a west coast flight that was over an hour late in taking off. Finally, about noon, Sawsilak called his office to let them know of the delay.

He was told that a plane had just crashed near Dulles. Concerned about the possibility of scaring people, he said to Gerstel, "I don't know if we should discuss this out loud, but . . ." at which moment Carolyn Leubsdorf, wife of AP reporter Carl Leubsdorf, came up and overheard what he was saying.

"Oh my God!" she said, "Carl took off about that time."

She immediately went to the check-in counter, where the clerk would not, or could not, give her any information as to which flight had crashed. Unfortunately for Mrs. Leubsdorf, she did not know which airline her husband had taken to the Midwest. She had a ticket for a later flight, because she and her husband feel that with five children, they should fly separately as often as possible. (That's the kind of caution that makes most air travelers laugh—except on days like December 1, 1974.)

Soon a few other people had learned of the crash, and the word spread quickly through the crowded gate areas. As usual when there are no details, many people assumed the worst, and within minutes there was loud crying and sobbing.

Senator Bayh and a few others moved through the crowd, talking to people, trying to calm them down. "Look," the senator said over and over again, "you don't even know what plane it was. And besides that, I was in a *small* plane crash twelve years ago, and I'm still here."

By 12:30, TWA had made no announcement regarding 514, with the exception of paging all those people who were waiting for it to arrive. In a lounge off the main lobby, it told those people that the flight had been diverted to Dulles. Later, it did announce that flight 427, 514's turn-around number, had been canceled. It did not say why—or how.

At 3:00 P.M., TWA provided a hastily-chartered D.C. Transit bus to take all the midwestern passengers to Baltimore-Washington International Airport, where the weather was less severe and planes were taking off with some regularity. The passengers, including the deaf girls, had to put their luggage on their laps for the one-hour trip.

Again, everyone had to wait. And still no one knew which plane had crashed or who was on it.

The scene in Baltimore was worse than that at National. TWA had only one gate available for all the passengers, and once it had processed all of them through the safety check point, it literally sealed off the area. No one was allowed to return to the main lobby.

Tempers had already flared during the safety check procedure. What bothered several people, in particular some of the reporters, was that the TWA security people insisted that because they did not have the newest detection equipment, each one of the deaf girls would have to unwrap each one of her Christmas presents. This so infuriated Rowland Evans that he shouted at the TWA man, "What the hell do you expect to find, a gun?"

When the very first present was opened, it contained a gun —a plastic toy one of the girls had bought for her brother.

When the group was still in Washington, TWA had suggested that the passengers might like to eat something, at the airline's expense, in the coffee shop or restaurant. Many people, familiar with airport food, declined the offer. Once they were behind the ropes in Baltimore, having waited for at least four hours, they were not too pleased to hear the announcement that there would be no meals served in flight.

When the flight carrying the deaf girls finally took off, the pilot tried to make amends for the less-than-polite treatment his deaf passengers had received on the ground. Unfortunately for the girls, he made his apology—on behalf of TWA—over the public address system.

By 6:00 P.M., Carolyn Leubsdorf was still trying to learn if her husband was safe. Finally, she took the suggestion of one of his colleagues and called the press room at the convention site. A friendly voice answered. "Carl?" he said, "Yeah, he's sitting right here, typing away like Woody Woodpecker."

For some reason, Carolyn Leubsdorf thought she should tell the clerk at the TWA counter that her husband, whom she'd been asking about for hours, was safe. The bone-weary clerk looked at her for a moment and then said, "That's great, lady. But my problem is, how am I going to replace that equipment?" He was referring to the airplane!

* * *

The small towns that dot the western slope of the Blue Ridge Mountains, out along Virginia Route 7, are not much more than an hour's drive from the Nation's Capital. Yet their pace and life-style is light years away. Even the mere forty miles to Dulles represents a chasm. But these are hardy people, and their response to tragedy is swift.

Volunteerism is traditional and taken very seriously in this part of the country. In fact, in the early afternoon of December 1, some of the volunteers probably remembered a similar search, years earlier, when another ill-fated plane went into the west side of Mount Weather at almost precisely the same spot.

By late afternoon, it was apparent to everyone on the

crash site that there were simply no survivors. Not a single body was taken to the hospital; instead, the abandoned schoolhouse in Bluemont, Virginia, was converted into an impromptu morgue. Bluemont, once a summer vacation spot for wealthy Washingtonians and now a blink-and-miss-it crossroads town, became a receiving house for the dead.

The greatest problem was not sustaining life. It was too late for that. The greatest problem was simply that of identifying the remains. One of the FBI agents at the crash site was shown a charred pistol holster. He said, "That looks like one of ours."

An investigator, perhaps thinking of the many relatives and friends of the victims, said that there was very little sign of shock on the faces of the deceased. Apparently there was no anticipation of the crash. Several days after the crash, a federal weather expert told a newspaperman, "We do know that drafts were occurring. When there are strong winds blowing on the ridges, you always get them. And if an airplane goes from a layer of air going one way to a layer of air going another, he is in for a surprise."

* * *

There is an immediate clash of interests following any airplane disaster. On one hand is the public's right to know, fueled not just by the press but by the intense need for reassurance that flying is truly as safe as the major airlines claim it is. On the other hand is the near-impossibility of determining within a short time after a crash exactly what caused it.

TWA 514, 1974's worst air disaster, was no exception. It would be days before the pieces of the puzzle could be fitted together. What's more, there were some unusual aspects.

The weather was, in a word, rotten. But 514 was not the only airplane in the vicinity at the time it crashed, and all the others made it.

Eastern Airlines Flight 878 was in a holding pattern over Dulles; later, its captain said that from the time he entered the overcast west of Richmond until he landed at Dulles, it was one of the most continuously rough flights he had flown in the last ten years. Similarly, severe turbulence was also reported by TWA Flight 99, American's 300, and United Airlines Flight 59, all of which made their approaches to Dulles at the same time as flight 514.

Another difference between 514 and the other flights is that for several minutes just prior to the crash, the Dulles Air Traffic Controller lost sight of 514 on his radarscope. When at 1109:54 he saw 514 reappear on his scope at an altitude of 2,000 feet, he tried to contact the airplane. He was going to warn the pilot that he was too low, but despite trying repeatedly for several minutes, he never could contact the plane.

Later, in a signed statement the controller explained that 514 may have disappeared from his scope because it had just entered an area of precipitation, which made the "target difficult to see."

It is possible that if 514 had not disappeared from the radarscope it would not have crashed. Still, even if the controller had noticed how low the plane was flying, his warning might well have been too late. Mount Weather stands 1,764 feet high. Flight 514 struck the western slope at an elevation of 1,669 feet.

Ninty-five percent of all crashes are put down as "pilot error." The rest break out between manufacturing defects, faulty maintenance, controller mistakes, and a few that are simply acts of God. But 514 was a compendium of adverse factors piling up one after another until the total "Load" imposed upon the flight crew proved to be more than it could handle. There were seven loads in all.

ONE: The weather was bad for flying, primarily because of the strong turbulent winds. As 514 descended, the rough air got worse, churned up by the Blue Ridge Mountains. Although the big Boeing was built to withstand such conditions, and the pilots were trained to fly in them, a bucking 727 is a handful. First Officer Kresheck was no doubt devoting his full concentration to controlling the aircraft. Furthermore, the fine print on an approach plate is ever so much more difficult to see in a bouncing cockpit.

TWO: The crew had not planned a flight to Dulles and was not familiar with the approach to runway 12. Had they used this approach on a regular basis, certainly one of the pilots would have remembered an obstacle more than seventeen hundred feet high.

THREE: The approach plate was poorly laid out, in the sense that it did not show a descent path in profile. Such a guideline, exhibited on many approach plates, would have shown the appropriate altitude for any given distance from the field. Mount Weather was shown on the chart, together with its elevation, but in very faint print. It is tragically clear that Captain Brock saw this information, but *he didn't believe it!*

FOUR: The radar controller issued an approach clearance quite a bit sooner than the crew probably expected it, for 514 was still about thirty miles from the field. Had the controller not turned the aircraft loose, he would have dictated its minimum altitude, and it would have cleared Mount Weather.

FIVE: The copilot was flying the airplane. Because he was always second in command, always flying from the right seat, First Officer Kresheck was in the *habit of obedience* to his Captain and not in the habit of challenging his judgments. Captain Brock determined the correct altitude from the "dumb sheet"—the approach plate—and then decided that the dumb sheet was wrong, and that his clearance allowed him to descend to 1800 feet. Neither the flight engineer nor the copilot said, "Hey, wait just a damn minute, let's not guess about a minimum altitude!"

SIX: Captain Brock was a victim of "polysemantics." When the radar controller uttered the words "cleared for the approach," he was imposing on the crew the responsibility of maintaining a safe altitude. Captain Brock very clearly thought that the responsibility had not shifted, and that the radar controller was authorizing him to descend to the initial approach altitude printed on the plate as 1800 feet. The controller was technically correct, but many other seasoned captains would have interpreted the clearance as Brock did.

SEVEN: The Boeing lacked a piece of equipment which might well have saved the day—a ground proximity warning indicator—which would have shown Mount Weather to be directly in the path of flight. The radio altimeter was working, but that reacted only to objects directly below the airplane, not those in front of it. The radio altimeter gave a warning, but it was too late.

In almost every single air crash the cause is eventually pinpointed. For example, the crash of the Eastern flight at Charlotte, North Carolina, last year was found to have been caused by the fact that the pilot and copilot were having a political discussion during final approach (However, ALPA, the airline pilots group, recently challenged the government's transcription of the tape, and asked that the investigation be reopened.); and the horrendous crash of the Turkish DR-10 outside Paris was attributed to the cargo's door having blown out. Thus one was caused by pilot error and the other by a manufacturing defect. But the true cause of the 514 disaster is not so easily ascribed.

According to John S. Yodice, the assistant secretary and Washington counsel of the Aircraft Owners and Pilots Association (AOPA), the question of what the pilot should have understood touched off "a raging debate. The evidence indicates that there is confusion. The FAA disagrees. The debate gets right to the heart of the air traffic control system and to the relative authority and responsibility of the pilot and the controller."

The details of this debate will be examined at length in a later chapter, but the basic issue is: does the controller have the responsibility of warning the pilot about altitude restrictions or is he just to provide a clear airspace (whether for takeoff or landing)? The Federal Aviation Administration (FAA) for whom the controllers work says it isn't the controller's job to specify altitude. The pilots, in general, feel that unless it is safe to descend immediately to the final approach altitude, the controller should not clear them for approach.

And there, for the moment, the debate rests. But it is interesting to note that since the Dulles crash, almost every pilots' manual—commercial, military, or private—has been changed to clarify this point. All of these manuals, incidentally, put the burden for terrain clearance on the pilot.

* * *

Near the middle of the construction maze between the Pentagon and National Airport sits the Quality Inn, Pentagon City. In contrast to the ever-changing roadways nearby, the inn is of a pleasant modern design, with more than ample parking space. On leaving the inn, the motorist drives past a sign that reads Happy Journey, Safe Return.

On Monday, January 27, 1975, that sign must have caused some people a moment's pause, for that was the day of the first public hearing into the crash of flight 514.

Air crash hearings are somewhat like fires: the greater the tragedy, the greater the crowd. The 514 hearing was no exception. The Capitol View Ballroom was packed. There were the members of the Board of Inquiry, the panel designated by the National Transportation Safety Board (a government body) to investigate the crash; its technical panel; some forty potential witnesses; a large group representing the press; and an equally large group of "interested parties," some of whom were lawyers representing the heirs of people who had died in the crash.

The tables in the ballroom were arranged in horseshoe fashion, with the witness table at the head, the technical panel on one side, and the parties on the other. Folding chairs took up the rest of the space in the large room, and as soon as anyone left the room, the empty seat was quickly filled.

The Honorable Louis M. Thayer, a member of the NTSB, was the board's chairman. (A retired admiral, he would later admit that prior to the 514 inquiry, the largest investigation he had handled had to do with a "jackknifed truck.") The hearing officer and investigator in charge was Rudolf Kapustin, an experienced investigator with NTSB's Bureau of Air Safety.

Among those testifying on the first day was Lloyd D. Brundage, a TWA captain who had been flying flight 99 on the day of the crash, and who landed at Dulles around noon.

Flight 99 was not cleared to land the first time it got near enough to Dulles, but was vectored to 6,000 feet and then finally cleared to a holding pattern at Blue Ridge, where it was held for half an hour. In no uncertain terms Captain Brundage testified that if he had been cleared the first time around, he would have immediately descended to 1800 feet. Like Captain Brock, Brundage was not familiar with runway 12. He told the inquiry that even though he flew into Dulles twelve to fifteen times a year, he had never flown that particular approach before.

When asked to explain what the words "cleared for the approach" meant to him, Captain Brundage said, "To me, 'cleared for the approach' means I can go to the final approach in the chart, in this case 1800 feet."

(An interesting note: Brundage testified that although the first officer was flying the plane that day, he took over the controls and made the landing.)

It is not the purpose of an NTSB inquiry to decide culpability, to place blame, but rather to determine exactly what caused the crash so that warnings can be promulgated in the hope of avoiding similar disasters. Yet, after the first day it was evident that the hearings were straying a bit afield. The questioning ran on for so long and became so tangential, that tempers grew short.

At one point, a reporter (Jules Bergman, the science commentator for ABC) was feeding questions to various panelists, hoping to get answers that would enable him to report at least *something* of substance regarding the hearings. There is very little love lost between aviation types and the press, caused by the unfortunate fact that it takes time—sometimes even months—to bring out the findings on a major air crash. And reporters are not used to waiting months to follow up on items of intense public interest. Their editors and superiors are probably even less patient, which may explain why a speaker at a recent controllers' meeting felt obliged to warn the audience about the press. "They'll go so far," he claimed, "as to wait till you go to lunch, and then rifle your desk!"

By the third day, the 514 hearings had deteriorated into a snarling mess. When it was all over, John Leyden, the president of the Professional Air Traffic Controllers Association (PATCO) told a reporter, "It's like a three-ring circus. The only thing missing is candy and popcorn."

Actually, by the end of the summer of 1975, almost ten months after the crash, the question of what the report would say had become academic. Apparently there was a rough draft of the final NTSB report, and—as usual in Washington—someone was leaking its contents. But few people expected the report to say anything other than the tragic fact that the pilot had not understood what the conroller meant when he said "cleared for the approach."

* * *

By mid-August 1975 all the lawsuits had been filed, all the paperwork was done, and most of the people who lost loved ones had somehow accepted the reality of the tragedy. The court case would take years, but eventually some formula would be used and, in ninty-two separate instances, the worth of an individual human life would be reduced to dollars and cents.

The Jeppesen approach plate showing the VOR DME approach to Dulles had been redrawn. Mount Weather, Richard Brock's "dumb reason," was properly outlined.

All during the summer of 1975, the few hardy souls who lived on the western slopes of the Blue Ridge noticed an unusual number of cars. But they weren't tourists getting a jump on the Bicentennial. They were ghouls. Souvenir hunters who wanted a piece of the tragedy. Something to take home with them to remind them of the year's worst air crash.

Today, the cars are still coming, slowing down to ask residents for directions to the site of the crash. Unfortunately, few of the ghouls are disappointed, for the area is littered with debris. In addition to pieces of the airplane itself, there are still shoes and belt buckles and other personal items to be found.

Some of the searchers appear disappointed not to find parts of the bodies. Perhaps they expected to because they read in the newspapers that more than two hundred sealed plastic bags of human remains were taken to the impromptu morgue at the Bluemont Elementary School.

The hill people just look at these ghouls and shake their heads. One of the most disappointed is the man who owns the slope acreage closest to the crash site. For all intents and purposes, it was on his land that the plane crashed. He had wanted to sell his property, a rather choice piece of land. But now he finds that no one wants it.

He shouldn't despair. Sooner or later, human nature being what it is, someone will come along and want to buy that property *because* of what happened there.

2

The Industry

Manufacturers make green airplanes. McDonnell-Douglas or Boeing or Lockheed each builds only the basic plane. The type and amount of equipment that a plane carries is chosen by the airline, and that choice involves everything from pillows to paint.

But when that airplane gets on the line, and is ready to carry passengers, all that has changed. The cabin of a large modern jet is filled with equipment, a dizzying array of dials and switches and buttons. The typical cockpit has more than 75 instruments in front of and above the pilots; more than 200 toggle switches, knobs, and levers; and close to a hundred lights indicating various functions. All in all, the devices and instruments total almost 700.

Among these are many that are optional, in the sense that neither the manufacturer nor the airline is required by federal law to install them. One such optional device is housed in a small gray box, usually mounted in the nose of the airplane, and connected to a tiny speaker in the cockpit. It's called a ground proximity warning system (GPWS) and by emitting a sweeping electronic beam some 2,500 feet below the airplane, it can warn the pilot that a solid object is in the way.

N54328, the Boeing 727 that flew into Mount Weather did not have a ground proximity warning device.

Why didn't it have one? There are a number of answers, beginning with the fact that many pilots are not convinced that the device is truly useful (they doubt that they would hear the taped warning, "Pull up!" in time for it to make any difference, and some worry that the device might malfunction and cause them to climb suddenly into the path of another airplane.)

Another reason is that in 1974 the FAA did not require collision avoidance devices in commercial airliners. But that answer is of small consolation to the families of the eighty-five people who died on the mountainside.

Resolution of the problem may stem from the lawsuits filed as a result of the 514 crash. One of the defendants in the lawsuit is Boeing, the manufacturer, who is being sued on the theory that because it knew all about collision avoidance devices, it was obliged to make such equipment mandatory. The company's defense will be: 1. the government didn't require it; and 2. the airline didn't see fit to install it.

As legal defenses go, the second is stronger than the first. But the lawyers for the plaintiffs are hopeful. They are well aware of the fact that if a jury finds *both* the manufacturer and the air carrier liable, the two will then fight one another to make sure they're protected against a similar verdict in the future. And that will mean a higher standard of safety for everyone concerned.

Civil liability, which is what we're talking about here, is always an effective form of regulation. (Without it, there wouldn't be many recalls in the automobile industry.) And manufacturer's consciousness, in all industries, is being sharpened (even those who are sanguine and greedy) by the lawyers who are telling them, "Your neck is out a mile, and we're going to tell the board of directors." That kind of blunt advice is overcoming the reluctance of numerous sales-oriented chief executives. It's a curious situation, but once the lawyers get the executive worrying about his own behind, the consumers usually benefit.

The conscientious man who builds an airplane says to himself, "I'm building a potentially dangerous instrument." And in terms of liability, it's very easy to convince a jury that an airplane is a potentially dangerous machine that should be built with extreme care. An auto manufacturer could say the same thing, but he rarely does—at least not publicly—because people feel that a car is only dangerous if it's driven fast or recklessly. Still, a brake failure at legal highway speeds is *damn* dangerous, and yet everybody drives a car. Thus, unlike airplanes, the mystique isn't there.

You'd have to search a long time before you found an airline passenger who said, "I'm not going to fly on X airlines because they use 727's [for example] and I don't think that's a safe airplane."

In my opinion, anyone who would say that is simply wrong. I feel that way because I've seen what goes into the design, testing, and building of an airplane (and, on a much more personal level, a helicopter.) There is nothing in American manufacturing today that even compares with the aircraft industry when it comes to building a safe and reliable vehicle.

The trusting airline passenger is protected against the horrible spectre of an unsafe airplane by two basic factors: the manufacturer's desire to build an airplane that gets a good reputation for ruggedness, performance, and comfort; and the certification process.

The safest of all American industries is probably the arms and munitions business, where a mistake in the plant could mean the end of the plant and everyone working there. That's a special case. Right after that is the aircraft industry.

The effort to provide ruggedness, performance, and comfort is governed entirely by the framework known as the certification process. Because no other form of transporation has this requirement—cars and boats are not certified by the government—an examination of it should help explain my contention that flying is the safest way to travel.

For the sake of discussion, let's say you want to build an airplane, either a brand new one or a modification of an existing design. What would you have to go through before you had a product you could sell to the airlines or the rest of the flying market? What you would have to go through is certification, which is broken down into the following stages: design, testing, and production.

On the design level, what you do is have the engineers sit down and work out what the new (or modified) airplane is going to do. For example, they'll ask, If we make the wing 40 feet long, how much will it lift? If the airplane weighs 16,000 pounds, and we want it to go 500 miles an hour at 40,000 feet, what sea level horsepower do we need to get it up, not just so that it can take off from a 5,000 foot runway, but to give it cruise speed at the desired altitude without burning too much fuel?

Given known technology, the engineers can make these predictions. Once they've made their calculations, they send their findings on to the management, and the next step then takes place—building the prototype or model.

The prototype is to be used for experimental flight tests, to see if the airplane, as designed, will do what it was intended to do.

Once the prototype is built, you send a test pilot up to wring it out. If a mistake has been made, the test pilot will find it, or it will find him—which is why the life insurance premium is so ungodly high for test pilots.

The pilot doesn't just test the plane to see if it can do what it's supposed to do; he purposely takes it to its outer limits to see if it will crack. These special tests are used to determine maximum speed, the maximum angle of bank and dive, and so on. While the pilot is flying these tests, special strain gauges are monitoring the effect on the various parts of the airplane. For example, is the aluminum-forming member of the airplane's skin holding up?

The whole idea is to push the prototype to the edges of flight parameters, or "the red line" beyond which sustained operation is dangerous. (Think, for lack of a better analogy, of what would happen to your car if your tachometer needle stayed in the red zone for hours and hours.)

Another type of test at this stage is for flight vibrations. This is of critical importance, because a series of parts (an engine turning at one speed, a generator turning at a different speed, and a cooling turbine way back in the fuselage turning at still another) vibrating separately can set up harmonics—a combination of vibrations strong enough to shake the whole airplane apart.

Day after day, in test after test, the pilot purposely exceeds the design limits of the airplane. If you want your airplane to be certified to cruise at an indicated airspeed of 300 miles an hour, you have to be able to show that it can fly indefinitely at 350. This is done to provide a safety edge, just in case the pilot gets in trouble and has to go over 300.

Another element of the testing process is to find out how much punishment the plane will suffer on a hard landing. In the helicopter business there's a chilling phrase that is used for this procedure. It's called the "drop test," where you just pick the ship up and drop it on the ground from increasingly greater heights—once again just to give it the edge.

Another type of testing is known as static testing, which is most definitely not done in the air. Taking a pressurized airplane as an example, you'd determine the pressure you want to carry in it—say, five pounds per square inch, which would enable you to fly at 25,000 feet without the cabin pressure ever going beyond 10,000, the point at which a human being begins to need oxygen—and pump it up double. Then you begin to cycle it: air in, air out, air in, air out; and you take it until it fails.

If you're trying to determine the amount of weight you want the fuselage to be able to carry, say 16,000 pounds, you double or triple it and just keep adding lead to see at what point the airplane buckles, begins to deviate from its normal shape. And then you watch to see if it snaps back when you remove the weight. Not too surprisingly, this is called destructive testing, which is almost as bad as it sounds, because you are deliberately breaking the airplane. Obviously this is never done in flight because it would mean killing somebody.

When all the data is complete, you call in the FAA, and they will selectively ask you to repeat anything they feel like seeing. The FAA will completely flight-test the airplane with you, and if they are satisfied they will certify the prototype by giving you what is known as a type certificate.

The next step is to go into production, but the FAA does not then fade out of the picture. Once you have a type certificate, you must then qualify for a production certificate. The FAA has certified the design and construction of the prototype and found it to be sound, but now they take a look at all the production machinery to make sure it is capable of duplicating the airplane faithfully.

There are two ways of doing this—of making sure the production doesn't deviate from the prototype to the extent that approval is doubtful. One is the presence of the DER, or designated engineering representative. This person is an employee of the company, but he is licensed by the FAA. He can "sign off," or approve, an individual plane, but in almost all cases, he cannot do so until the FAA is thoroughly convinced that the line is functioning properly. The FAA may have approved the first six or seven planes off the line before it transfers its function to the DER (who cannot sign off a prototype or the first production unit).

The FAA follows this rigid procedure because the first run of a production line may well turn out a less-than-uniform product (because new parts may be involved, and other changes are taking place aimed at increasing efficiency).

The other FAA requirement that governs faithfulness of production is that every manufacturer must have a quality control department. This of itself is not unusual, but the FAA puts teeth in the process by making the manager of the quality control department directly answerable, by law, to the chief executive officer of the company.

The reason for this is that if the manager gets, say, a number of defective wing members, which could necessitate shutting down the entire production line, he cannot make the decision to keep the line running. That decision must come from the top man in the company. It's possible that a department head wanting to meet his quota might let something marginal slip through, thinking, "Well, it'll never break. We've got so much extra margin built in." But the president of the company wouldn't take that chance, because he might lose his job if he got caught. The other guy, bucking for promotion, just might be willing to run the risk. This is a procedural safeguard that's built-in to the aircraft manufacturing business.

What these examples mean is that safety is the prime concern from the very beginning of the process. It's all very detailed and safe, which is why most of the airplanes that are put in the air never, never have a structural failure, unless they somehow get into conditions that are way beyond the design limits of the aircraft. They simply don't fall apart in the air, and this is as true of Piper Cubs as it is of jumbo jets.

There is nothing even remotely comparable to this in the automobile industry, although, in fairness, it should be noted that while hundreds of cars come off the assembly lines each hour, the top production in the aircraft industry is that of the Cessna Skyhawk—eight ships a day.

The tone of the FAA bureaucracy is, if in doubt, tell the manufacturer to redesign or rebuild. Or, the agency will run new, even more stringent tests to convince themselves that the plane or part in question will hold up. Not only does this make the product safe, it makes it damn expensive to build. (People wondered why the SST carried a price tag in the billions. For one thing, it had millions of miles of wiring, all of which had to be checked and certified every inch of the way. And its engines cost almost $3 million apiece. Even the engines in my six-passenger AeroCommander are $100,000 each. It all adds up.)

Small airplane companies are put out of business all the time because the FAA insists on a change that they cannot afford to make. Certification failure is not uncommon, and production certificates can be and are revoked, in some cases because the company didn't have sufficiently qualified people to run the production.

All of this testing and checking is done to make sure that the airplane is safe before passenger number one sets a foot on the boarding ramp. Throughout the entire manufacturing process, the FAA stands up for the user—the ultimate consumer of the aircraft product—and it doesn't just look over the manufacturer's shoulder, it steps right in front of him to make sure the job is being done properly. One of the reasons for this is that the aircraft industry is very competitive. And whenever there is competition, watchfulness is important.

I'm not suggesting that aircraft manufacturers would be cutting corners left and right if the FAA weren't watching. They wouldn't because in the long run they couldn't afford to. But the competition for buyers is keen, and that means that compromises have to be made in the design and construction of airplanes.

For example, there's always a potential compromise between comfort and performance. (This does not involve safety standards; they're going to be met or the airplane will never get off the ground.) The design of any airplane, from a two-seater to a flying lounge, necessitates some form of compromise involving the three main factors—comfort, performance, and economy. The mix of these factors determines the characteristics of the final product.

An example of a compromise in favor of performance, especially in regard to speed, would be the Convair 880. Until the jumbo jets came out, the fastest airliner in standard use was the 880. It's cruising speed was 550 to 600 knots, faster than the other two planes in the initial group of airliners, the DC-8 and the 707.

TWA and Delta bought a lot of 880's, mainly for their speed. But in gaining the speed, they lost on other fronts. The 880 had a short range, was quite loud, and burned a lot of fuel. Unlike the other four-engine planes, they couldn't go coast to coast because they couldn't carry enough fuel. Their approach speeds prior to landing were higher, which made them a problem for the air traffic controllers, who had to allow a big hole to get an 880 down.

So the compromise meant that in opting for speed, the company settled for less in the way of comfort and economy.

This kind of competition is far more apparent in the field of general aviation, which will be discussed in a later chapter; although it exists in the air carrier business, the passenger is blissfully unaware of it. During the design phase, the engineers spend weeks and months trying to determine what top rate of speed to give the plane, how to make it able to climb swiftly out of bad weather, and any number of other factors, all of which (they hope) will make theirs a popular airplane.

The passenger, however, neither knows nor really cares about this kind of competition. The average American passenger does not decide what kind of airplane he's going to ride on, with the possible exception of taking a 747 rather than a 707 so he can leave a half hour later for a flight to the opposite coast. His only real decision has to do with time.

Almost all the planes of the various major airlines have the same size seats (in coach and first class) and they all go about the same speed (where speed is even a factor—on short hauls it makes no difference whether you get an Electra or a DC-9). The differences between a Boeing 707 and a DC-8, the classic original jets, are very subtle. Passengers simply do not call the airlines and say, "What kind of equipment is that?" Of far greater influence is the kind of meal to be served, the configuration of the seat (If it isn't too full will I get the simulated first-class seat in coach?), and perhaps the choice of movie.

The person who buys the airplanes for the air carrier companies does not have to worry too much about whether or not his customers will prefer one brand over another. Almost all the airplanes that the passengers have, theoretically, to choose from are comfortable, and can be equipped with nice paint jobs and nice galleys and nice stewards and stewardesses. The guy who buys the planes from Douglas or Boeing is really most interested in how many people it will carry. He's buying the way bus companies do. Since comfort is built in and performance close to equal, he's thinking in terms of economy: "What route am I going to put it on? What short fields do I have to go into? How practical is it from an economic standpoint?"

In the past few years there have been some slight changes in this picture. There's been some "romance buying," where companies put jets on routes where jets aren't really justified (but another company had done it first so the others felt compelled to). And most everyone is aware that for the first time in fifteen years many of the companies have adopted new paint schemes, in a belated admission that "looks" help sell tickets too. Also, some manufacturers have begun to take out ads in the popular magazines, in what appears to be an attempt to create passenger preference for, say, a Douglas over a Boeing. This probably has little or no effect on the public, but may have on the buyer for Eastern or American or TWA, who sees the ad and says, "This manufacturer is making himself popular with the public, so I'll buy his product and the public will accept it."

Still, the picture remains much the same. The average American passenger has little or nothing to say about the type of plane he flies in. Even with the Eastern cattle cars—the shuttle service between New York and Washington, where the customer is guaranteed a seat as long as he's willing to wait—passenger preference plays no real part. The only real option the passenger has is to wait for a type of plane more to his liking. And few people bother to do that.

The passenger doesn't have to make a choice because the government, through the FAA, assures him that the airplane is safe. But it's worth remembering that the government is in the background once the aircraft is certified. From that point on, the passenger relies chiefly on the reputation of the company and the airline (which checks the condition of a plane before and after each flight).

There is another safety factor that should be discussed, and that is the concept of the manufacturer's conscience.

Not all companies are alike. Going outside the field of aviation, consider the two examples of ITT and Xerox. ITT has a corporate philosophy that is so highly competitive and cutthroat that its executives are burned up like air traffic controllers of old. Xerox, on the other hand, while a smaller company than ITT but still very large, has a tremendous perception of the end-user. Without being told, or challenged, by the Federal Trade Commission, it will go out and correct something that it views as less than desirable, even though it knows it could probably get away with it, and save some money. The reasoning is that to do so will build up owner-user loyalty, a distinct plus in the long run. Xerox feels a responsibility toward the consumer, who has so little to say about what takes place at the manufacturing end.

In the aviation manufacturing business, the perception of and concern for the end-user is less visible but nonetheless real. If an accident occurs in the field, it can ground an entire fleet of airplanes until the fault is discovered and remedied, a process that may involve putting everyone in the plant on triple time, grinding out replacement parts, and furnishing warranty (free) labor to install them. The manufacturers face the hard fact that if one of those three hundred ships breaks and the others *might*, the FAA may say, "Fix 'em all."

Under federal law, there are procedures that supposedly guarantee the passenger that the first time a defect is discovered in an airplane that has already been certified, an industry-wide bulletin will go out that makes a corrective fix mandatory.

I say "supposedly," because there have been some notable exceptions, with horrendous results.

The public should know that there is a constant battle going on in most companies between engineering and sales. The engineers say, "Fix the damn thing." And sales replies, "Hey, wait a minute. We're bidding against the Lockheed 1011 with six foreign governments, and we announce we've got a defective design? Let's cool it, and maybe it'll never happen again."

This is the kind of tug-of-war that once produced a judgment by a jury against Beechcraft for $2 million in damages —and $20 million to punish them. The company knew it had a defective fuel system—if you turned onto a runway at a high rate of speed it would quit at takeoff—and it didn't tell anybody.

The worst example of this continuing problem was the cargo doors on the Douglas DC-10.

As I've said, aircraft design and manufacture is quite sound, and the certification process is thorough. The most serious problem arises when a defect somehow slips through the certification stage and crops up in an operating aircraft.

When that happens, the FAA and the NTSB are immediately on the scene. They investigate, talk with company officials and the airplane's crew, and decide what type of action the FAA should take. It can issue an airworthiness directive (AD), its harshest remedy, which means, "You *must* fix," and that entails grounding all planes of the same type and shutting down production of new ones until the defective part has been repaired. Or, the FAA can take several less stringent measures, such as the "voluntary fix," which all but leaves the problem up to the company's own discretion.

And that's what happened with the DC-10. The result was the worst single disaster in aviation history, a tragedy made all the more horrible by the fact that the company knew what part might fail—because it had failed before.

The tragic chain of events began during the certification period.

Airplane manufacturers seldom make all the parts for the airplane. In the case of the DC-10, the contract for building the fuselage and the three cargo doors was let by McDonnell-Douglas to General Dynamics, with the actual work being done by its Convair Division. In May of 1969, while testing to see what would happen if certain parts did not function properly, Convair reported in a memorandum that if the plane took off without the cargo doors being properly fastened, the result could be calamitous. According to the memo, "Door will fully open—resulting in sudden depressurization and possible structural failure of floor. Also, damage to empennage by expelled cargo and detached door."

What worried the engineers was the possibility that if the door blew, the main tail controls could be seriously, perhaps even totally, damaged because their cables ran directly through the part of the floor that would probably buckle if the door came open.

But the engineers did not include this last concern in their memo, and the company did not think the doors would ever be a serious problem. For one thing, locking the doors was a three-step process (two of the steps involved pressing buttons so that electrical motors did the heavy hinging work; the third step was to close the handle, which was designed so that if the first two steps had not worked, the handle could not be closed in a flush position), and for another, there was a light on the dashboard to warn the crew if the cargo doors weren't fully closed.

In May of 1970, a year after the memo had been written, the prophecy came true when a forward cargo door blew during ground pressurization tests. The result was a small explosion, which bent the floor structure. The company then redesigned the doors, adding a small vent door as a further safety feature. Some of the engineers, however, would later refer to this alteration as "a Band-Aid fix."

The plane was certified (without the FAA ever being told of the 1969 memo) and the company was satisfied that it had a triple-fail-safe procedure: the vent door would close only after the main cargo door was properly closed, and neither would close unless the handle was flush with the plane's skin; also, if the vent door did not close properly, the plane could not be pressurized, and any pilot on learning this in the air would immediately land, the door would then be properly closed, and he could take off with everyone safely protected.

The airplane was certified on July 29, 1971, and put into service throughout the country six days later.

During its first ten months of service, the DC-10 was a relatively trouble-free airplane, with one exception—the cargo doors. Several airlines reported problems. Apparently it was possible for a baggage agent or ground crew member to force the handle flush, even though the doors had not actually closed. And it didn't take a Samson to do it. Still, only company service bulletins were issued recommending proper closing procedures.

One June 12, 1972, a day on which I happened to be in the Detroit Metro Airport, an American Airlines flight left Los Angeles bound for New York. Stops were scheduled at Detroit and also at Buffalo. The airplane was a DC-10.

After leaving Detroit under ideal weather conditions, the captain switched off the no smoking sign and the passengers were relaxing as the craft flew over Windsor, Ontario, heading for Buffalo.

Suddenly, there was a muffled, boom-like sound, loud enough for all the passengers to hear. Within moments, a gray fog began to filter through the airplane. Up in the first class section in the fourth row, an aisle floor hatch flew off and struck a passenger in the face.

There was screaming throughout the plane. Three stewardesses, one standing and the others seated, were thrown into the aisle. Sections of the cabin floor collapsed, and a stand-up bar was torn from its moorings.

The aft cargo door had blown, causing explosive decompression which in turn buckled the floor above the cargo section and severed numerous cables that were connected to vital instruments and controls.

The crew responded magnificently, calming people down and getting everyone to assume emergency landing positions. The pilot, flying only with the trim-tabs—an extremely difficult task—was able to bring the plane around and land again at Detroit.

Miraculously, no one was seriously hurt and no aluminum got scratched, which meant that there was not a great deal of publicity. But the McDonnell-Douglas Aircraft Company knew it was a devastating discovery. In theory, this kind of accident *could* happen with every DC-10 they had sold or would sell in the future.

The FAA and the NTSB immediately began to investigate the accident. And at this point the picture began to get murky.

The Los Angeles Regional office of the FAA wanted to issue an immediate AD (the toughest legal tool in the agency's kit), and in fact prepared a draft for an AD, in telegram form, which was to be held until the exact reason for the accident was known.

The AD never went out. Within days, then FAA Administrator John H. Shaffer, placed a call to the president of McDonnell-Douglas, Jack McGowan, and asked him what the company had found out in regard to the cargo door problem in Detroit. McGowan told the administrator that the trouble was caused by insufficiently heavy wiring which kept the doors from locking properly, and that the company could fix it without the necessity of an AD.

Shaffer bought the businessman's story, they made a Gentlemen's Agreement, and Douglas "solved" the problem by requiring—through a series of service bulletins—the installation of a small round window in each door so that the person closing it could see whether or not it was firmly shut.

McGowan called the head of the Los Angeles Regional Office, relayed to him the agreement he had made with Shaffer, and the Airworthiness Directive was killed.

FEAR OF FLYING [31]

And so, twenty months later, were 346 people, when a DC-10 owned by Turkish Airlines lost a cargo door over the Orly Airport, near Paris. It took less than two minutes for the explosion to collapse the floor, destroy the controls, and send the passengers and crew to the ground in history's worst air crash.

The investigation into that crash taught the company what it already knew, or should have known, that despite its "Band-Aid fixes," a moderately strong person could still force the handle closed. The auxiliary instructions printed in English made no difference to the French baggage handler who closed the door. He couldn't read that language.

Seventy-two hours after the Paris crash, the FAA issued an AD calling for the immediate repair of cargo doors on the DC-10.

* * *

As I said earlier in this chapter, I think the FAA has an enviable record in so far as its certification process is concerned. I also think that McDonnell-Douglas is a classy and usually cautious company. But once an airplane has been put on the line, scores of huge birds sold at a price tag of $25 million apiece, the economic pressure to resist an Airworthiness Directive is very great.

Subtle pressure was also brought to bear, for in this case the top government man involved had a history of being a buddy of the major manufacturers.

The extent of this closeness between the FAA and the industry (which one critic has labeled incest) will be examined in a later chapter. For the record, it is ironic that when John Shaffer recently left government service, he signed on as a board member of Beechcraft.

The worst tragedy in the history of manned flight may have been caused by a corporation's desire to save money.

And the worst air disaster of the year 1974 may have been caused for the very same reason. But in the case of TWA 514, there is a terrible difference.

Recalling all the DC-10's in use and shutting down the production line would have meant an immediate financial loss of millions of dollars a day. In stark contrast, a ground proximity warning device might have saved flight 514: at the time of that crash, only one company had a finished and FAA-certified device (Sunstrand Data Control, Inc., of Redmont, Washington) at a price of $6,225. On a 727, 6,200 bucks wouldn't even pay for the Muzak system.

3

The Government

"It has been noted before that regulatory agencies, like people, sometimes suffer from hardening of the arteries with advancing age. Symptoms of such a process have been noted in the FAA. Administrative delay and inactivity is bad in any agency; in the case of the FAA, it may literally endanger human life. Instances of completely inappropriate bureaucratic slowness to act, and inaction, are noted throughout this report.

The subcommittee found throughout its inquiry—from the DC-10 crash to its most recent investigation into the feasibility of requiring Ground Proximity Warning Systems— a tendency for the agency to avoid the role of leadership in advancing air safety which the Congress intended it to assume. This is manifested primarily by the FAA's willingness to let the industry engage in self-regulation when vital safety measures are concerned. In some instances. this abdication of responsibility has been coupled with an administrative lethargy—a sluggishness which at times approaches an attitude of indifference to public safety. This must stop."

> *"Air Safety: a Selected Review of FAA Performance" Report by the Special Subcommittee on Investigations of the Committee on Interstate and Foreign Commerce, U.S. House of Representatives. January 1975*

Every day of every year, congressional committees and subcommittees are busily churning out reports. In too many instances, by the time hearings have been held, testimony heard and digested, and the report written and printed, the cause is no longer fresh and the public no longer interested.

Once in a while, however, a report comes out that warns of a continuing danger. The report cited above is a classic example of a most timely cautionary tale. The shame is that so few people have read it—and that for the most part its recommendations have been ignored.

On the day after the crash of the DC-10 outside of Paris, the worst crash in history, the Subcommittee began its investigation of certain FAA policies and problems. The five-member Subcommittee (chaired by Democrat Harley Staggers of West Virginia) had a staff of nine, heavy on the investigative experience side with three of its members former FBI agents.

Its work centered on seven specific areas: the cargo door problems with the DC-10's; the ground spoilers on the DC-8's; the Ground Proximity Warning Systems; the DC-10's difficulties with its CF-6 engines; certain FAA policies regarding the issuance of Airworthiness Directives (AD's); FAA payola; and the transport of hazardous materials. A series of seven public hearings were held between March and September of 1974.

It was hard work, made even harder by the FAA's less-than-enthusiastic cooperation. Congressional investigators are not without power of their own, however, and the staff decided that it would be helpful to have certain FAA documents. It subpoenaed them and was surprised to find, among other items, an internal memo outlining the history of the Gentlemen's Agreement between FAA Administrator Shaffer and the head of McDonell-Douglas.

Two members of the staff, Ben M. Smethurst and Mark Raabe, both former FBI agents, had spent a great deal of time studyimg the GPWS. It was, so to speak, their baby.

On Sunday, December 1, 1974, Smethurst was relaxing in his Springfield, Virginia, home, a short drive from Washington, D.C. He'd worked hard on the report and was pleased that it was finally ready for printing. Suddenly, the radio announcer interrupted the scheduled program to say that a plane had crashed near Upperville, Virginia. Smethurst had a horrible feeling that his report was not quite finished.

Within an hour he was on his way to the crash site, where he met officials of the National Transportation Safety Board. He walked with them as the bodies were removed and a wooden stake driven into the ground to indicate where a corpse had been found. The incredible carnage was uppermost in his mind, but every once in a while he was struck by the thought that the TWA plane had not been equipped with what he'd come to refer to as a "ground prox" device.

Smethurst's colleague Raabe was equally shaken, first by the news of the crash and then by an experience that entailed viewing the wreckage from a different perspective.

On December 3 when the rain but not the wind had let up, Raabe and Daniel Manelli, the Subcommittee's chief counsel, were the only passengers on the Sundstrand Corporation's Beechcraft King Air E90. The purpose of the flight was to duplicate the approach flown by TWA 514 in a plane that *was* equipped with a ground proximity warning device.

The plane took off from National Airport in the late afternoon, and there was plenty of daylight left by the time it reached the foothills of the Blue Ridge Mountains. As Raabe said later, "Finding the crash site was easy. You just looked down until you saw a slash taken out the mountain. The plane had 'ravined' it. There were the remains of the airplane scattered everywhere, and you could see the rescue crew still working. The first sight of it was simply a great shock.

"From viewing the crater that the plane had made when it struck, we could figure out the direction it had been flying. And we already had preliminary information as to the altitude. Our pilot headed west at 1,700 feet and we made several passes."

Flying at 180 knots, 514's airspeed at the time of the crash, the three men waited to see how soon the GPWS would sound a warning. Raabe, standing just behind the pilot, and Manelli, in the copilot's seat, held stopwatches. As they flew on, the mountain loomed larger and larger, filling the plane's windows.

Suddenly a great squawking sound erupted in the cockpit. "PULL UP! PULL UP!" The plane was nearing a ridge, well ahead of the crash site, and the elevation triggered the warning device. The message was uttered, said Raabe, "In this loud voice I'll never forget."

Just past the ridge the device went silent, but it began to scream again as the plane approached Mount Weather. Raabe said that seeing it come up on them was like "sitting in the front row of a movie theater."

After several more passes the men compared notes and times. The device had gone off initially at fourteen seconds before the plane would have struck the mountain at the same spot as the Boeing 727. The next warning sounded at seven seconds.

All three men in the airplane, Raabe, Manelli, and the Sundstrand pilot, were convinced that even seven seconds would have been enough of a warning for TWA 514.

* * *

The federal government has had a hand, and usually a helping hand, in the growth of aviation since the earliest days. In fact, some pilots will tell you that only the Wright brothers were free from government regulations.

In 1918 airmail service between Washington, D. C., and New York City was officially begun, and although few if any people knew it at the time, the seed had been planted for the growth of the major airlines of America.

For the record, that first flight was hardly a success. The trappings were impressive, though. Three military planes, Jennies, were brought to a small polo field not far from the Capitol by the Army and assembled on the spot. While President and Mrs. Wilson, plus scores of lesser dignitaries, milled around the field, mechanics struggled for forty-five minutes to start the first plane. Finally, someone remembered to check the fuel tanks. Empty. Once airborn, the first mail plane buzzed the field a few times and then lit out for New York. Several hours later, it ran out of gas and had to put down in a field. The next day a postal truck picked up the specially-stamped mail and brought it *back* to Washington. (One should resist the urge to view this as symbolic.)

For the next twenty years, a government contract to fly the mail kept the wolf from aviation's door. It was a freewheeling exciting time for the pioneers as they taught the nation that the Wright brothers were not idle dreamers. Many older Americans have fond memories of craning their necks, looking skyward, and seeing a Ford Trimotor pass overhead "carrying the mail."

By the mid-'20s, faith in the future of the airline business was so strong that there were forty-four companies that could be called airlines. Granted, some of them were shoestring operations barely able to keep up the payments on their typewriters, not to mention their planes. But it was from this group of early companies that, through merger or acquisition, the modern airline giants came into being. And since no one was making money just carrying passengers, it was the airmail contract with the government that kept these companies going.

Recognizing both the growth and the potential of aviation, Congress passed legislation in 1926 establishing the Bureau of Air Commerce. From then until 1938 the government was, as one writer on aviation put it, "a passive participant" in aviation matters, as opposed to being a true regulator. It began to operate and equip airports, but only in a limited way.

In 1938, however, the picture changed with the creation of the Civil Aeronautics Administration, a new government agency set up under the Federal Aviation Act of 1938. Now the government took an active role in such matters as the training of civilian pilots, air traffic control, and the operation of airport control towers.

Within a few years America was deeply involved in another world war, and the influence of the military became the predominant factor in the entire aviation industry. But after the war, the CAA reestablished its control. In 1946 the small agency had grown to a work force of more than ten thousand. That, however, was just the beginning.

As old Washington-watchers will tell you, the bureaucracy is like nature in abhorring a vacuum, and it did not take the CAA long to fill its vacuum with new posts and responsibilities.

When the 1938 act was rewritten in 1958, the old Civilian Aeronautics Administration became the Federal Aviation Administration, a young giant of a governmental body employing some 29,000 people. Since then the FAA has "just growed" in the classic federal fashion. Today it employs more than 55,000 people and has an annual budget of $2 *billion*. It owns and operates more planes than some airlines.

There is hardly a place on earth or in the sky that is free from the regulatory power of the FAA. To the average airline passenger, the FAA is that unseen force that insists the stewardess demonstrate how to use the oxygen mask—or demands that all briefcases be placed under the seat and not in the overhead rack. (In his book *Airline Safety Is a Myth* Captain Vernon W. Lowell retells the classic story of the stewardess who'd had it with a passenger who insisted on stowing his briefcase in the rack; finally, out of patience, she said, "Sir, there are only two places where I'll let you put that briefcase, and one of them is under the seat.")

Lyndon Johnson used to be fond of quoting the maxim, "A man's reach should exceed his grasp." Apparently the FAA took the former President literally, for its reach is seemingly endless. The FAA now monitors the construction and testing of all aircraft; it regulates the training and performance of all pilots, from the most senior airline captain in a 747 to the suburban housewife who rents a Cessna 150 every Sunday afternoon when the weather is clear; it employs, trains, and supervises the air traffic controllers; it tells manufacturers and airlines what new equipment (such as the Ground Proximity Warning System) must be installed in an airplane or helicopter; it tells airports how to operate, down to X-raying you and your luggage. And that's only a partial list.

Indicative of the power of the modern FAA is the fact that after the Department of Transportation's investigative arm, the National Transporation Safety Board, reports its findings as to the cause of an air crash, the FAA is under no statutory compunction to take remedial action. The safety board is an investigative body that determines causes. It can make recommendations, but the FAA, and the FAA alone, decides whether or not to follow them. That's power.

My point in all this is that the FAA is charged primarily with ensuring air safety, and because of the agency's record in that regard, one has to ask if all of this regulatory power is truly exercised toward that end.

* * *

All federal agencies have their detractors, but the FAA seems to bring out the beast in people. In a much-talked-about article in the July 1975 issue of *Playboy* magazine, staff writer and student pilot Laurence Gonzales rapped the FAA. He claimed the agency "has the problem of simultaneously trying to promote a very profitable airline industry and safe flying—things that are beginning to seem mutually exclusive." The author also charged that "the FAA doesn't seem to know how to order its priorities."

That last charge echoes one made by Captain X in his recently reissued book, *Safety Last:* "Many governing bodies in Washington know their limitations—the Federal Aviation

Administration is not one of them." I once appeared on a television talk show with Captain X—a senior captain for a major airline—and know his concern to be well-founded and genuine. One of the things that bothered him most was former FAA administrator John Shaffer's conduct in regard to the air traffic controllers' "sick-out" in March of 1970 (a subject I'll return to later) and his generally poor record. When his book first appeared in 1972, X wrote, "FAA administrator John H. Shaffer, who rules a vast empire of 48,000 supposedly safety-conscientious employees, is one of the worst offenders."

In his book on air safety, Vernon W. Lowell, also mentioned the great increase in the number of employees within the agency. But, he said, "Sheer numbers do not guarantee efficiency. All too many pilots feel that the FAA has degenerated into a bureaucracy which often engages in the face-saving of its public image rather than the pursuit of air safety."

Recent years have seen even more harsh criticism of the agency. In 1972 Philip I. Ryther wrote a scathing attack on the FAA under the title, *Who's Watching the Airways? The Dangerous Games of the FAA*. What made Ryther's book so startling was that its author was a former FAA official. If the title was not sufficient indication of the book's point of view, the author wasted no time in stating his position; in the introduction he wrote:

> This book presents a case by case indictment of the top management of the FAA. Much of the material is gathered from still-secret FAA reports and from letters and other internal documents. All of the incidents in this book are true. All can be documented. Together they show that through its incredibly poor management the FAA has failed to afford the American public the protection it so badly needs in the field of aviation safety.

Mr. Ryther's case is also highly personal: he claims that by pushing for increased safety measures (in a study that came to be known as the Ryther Report, even though it was never officially adopted or promulgated) he succeeded only in bringing about his own forced retirement from the FAA. His carefully documented book makes it very hard to disagree with him.

In 1958 Ryther, for seven years a budget officer at the Pentagon in the office of the comptroller of the Secretary of Defense, was working in California. He'd been enticed away from the government by his former boss and was the director of financial controls for Northrup Aviation's Van Nuys division.

He was lured back as a management analyst with a government rating of GS-15 (one step below the management level) in February 1959, when the legislation creating the FAA went into effect.

General Quesada, the first head of the FAA, was shaking up the place by bringing a number of new people aboard, people who had not worked for the Civil Aeronautics Administration. Within a short time Ryther found himself up against the Old Guard. "The FAA, I was beginning to find, was, as a House committee later concluded, an organization with more independent empires than medieval Europe."

After spending a year helping with the effort to reorganize the myriad departments within the agency, Ryther was assigned to a project that brought about his first real conflict with the top brass.

By 1961 the FAA had a new administrator, Najeeb Halaby, who recognized that there was a problem in the area of supply—the buying, storing, and allocating of various government materials to FAA sites around the country. Ryther and several others were told to study the problem, and they found a less-than-ideal situation. Using Western Electric as a model because its function in relation to AT&T closely paralleled that of the FAA, they discovered that while it took Western Electric one day to process a request for equipment, it took the FAA personnel an average of ten days—seven days "when we were in a special hurry, and reached one day processing only when we needed the supplies in an emergency."

One of the most startling findings was that it cost Western Electric $4.76 to process every one hundred dollars' worth of supplies; it cost the FAA, in taxpayers' dollars, $17.95.

Ryther recommended to Halaby that the agency hire an outside consulting firm to study the problem further and to come up with suggestions for streamlining the terribly outmoded FAA supply system. (For one thing, there was no central office for procurement and disbursement; everything was done individually by the separate installations.)

At Ryther's suggestion the agency retained Harbridge House, a consulting firm in Cambridge, Massachusetts. Its report was, in Ryther's words, "a stinger," calling for significant changes in the FAA way of doing things. Some of the recommendations were eventually adopted, some weren't, but Ryther became known to the Old Guard as the man who had blown the whistle. He didn't know it, but his days were numbered.

Ryther's history over the next few years reads like a primer for the Ernest Fitzgerald case (the Pentagon civilian employee who exposed the massive cost overruns on the C-5A transport plane). Fitzgerald eventually sued and got his job back. Philip Ryther was not as fortunate.

In 1969 Ryther was given the assignment of studying the FAA's Flight Standards Service. A year later he submitted the report of his team, and among the thirty-one recommendations were two that proved to be highly significant, even prophetic. One dealt with the number of flying schools around the country that were not certified by the FAA. The investigators thought this represented a potentially dangerous situation, one that in extreme cases might be similar to the blind leading the blind.

The other prescient recommendation had to do with the fact that certain charter airlines that carried passengers were allowed to employ pilots whose experience and qualifications were nowhere near those required by the commercial airlines. The study team's recommendations called for beefing up these requirements if the for-hire lines were to continue carrying human cargo.

The Ryther Report was never officially adopted or even approved by the FAA.

The report was submitted on March 30, 1970, and on October 2 of the same year, a two-engine Martin 404 carrying the Wichita State University football team crashed in the Central Colorado Rocky Mountains.

The pilot had been directed to follow U.S. Highway 6, which snakes its way through a canyon ridged by peaks ranging up to 13,000 feet in height. Observers—startled observers—later testified that the plane seemed to be as low as 500 to 1,000 feet. Unwittingly, the aircraft flew past the climb-out point. When it got to the end of the canyon, it was trapped, and could not clear a mountain that formed the back wall of what had turned out to be a box canyon.

An NTSB investigation revealed that the crash was exactly the kind of tragedy that Ryther and his team had warned against in their stillborn report. It went this way: the plane was owned by a charter airline, Golden Eagle, whose president was the copilot; the pilot was a mechanic for Golden Eagle who occasionally doubled as a pilot. The charter company president had decided to disregard the official flight plan and give his passengers a scenic view of the canyon floor. The decision to do so was in direct violation of various FAA regulations, but the pilot apparently did not see fit to argue with the man who signed his paycheck.

When the plane crashed into the side of the mountain and burst into flames, the trees provided a partial cushion which saved some lives.

Of the 40 people on board, 30 died and 10 survived. One of the survivors was the man who owned the airline. His mechanic-pilot was killed.

The news of the crash stunned Philip Ryther, the man who had all but predicted it.

Eventually the FAA would act on what had been recommended in the Ryther Report and would change the standards so that charter flights could not be piloted by unqualified or inexperienced people. But it was not done by implementing the report, one of the many reasons why Ryther found himself becoming, as he put it, "persona non grata."

Determined to see that some of the other ills he and his six-man investigative team had uncovered were attended to, Ryther began to go over the heads of the middle management people within the FAA. And that is just not done in Washington. Following a long series of Kafkaesque happenings, Ryther agreed to resign from the agency in December 1970. After twenty-six years with the government, eleven of them with the FAA, it was, as he said in his 1972 book, "an unceremonious and frustrating end."

In 1971, prior to writing his book, Ryther made various attempts to get people to listen to his story and to do something about the bureaucratic lethargy in the higher ranks of the FAA. He visited senators and congressmen who promised to help but didn't. In March 1971 *Look* magazine ran an account of Mr. Ryther's frustrated efforts. Armed with that backing he approached OMB, the Office of Management and Budget, which oversees the various agencies, and when that did little or no good he even contacted the White House. That did not bring the kind of response he'd hoped for, so Ryther sat down and on July 9, 1971, wrote directly to President Nixon. This time he got a response—perhaps because he'd sent copies to the press and to sixteen members of Congress.

A White House aide called him and said that a special task force was being set up to study his charges, and that it would be headed by John Dean, a young counsel to the President who would later make history. But that never happened. Six weeks later he called that same White House aide again and was told that the matter had been referred to John Ehrlichman's domain, where it was being handled by Egil Krogh, Jr.

Ryther called Krogh but never got through to him. He was told that someone else was now in charge of his case. He called that person, left his name, and never heard from him—or the White House—again.

When his book appeared in 1972 (an "as told to" with Stephen Aug of the Washington *Star*, a specialist in covering the regulatory agencies) it was well-reviewed, but it did not spark any new investigations into his many charges.

Philip Ryther had given it his best shot. He had gone up against the bureaucrats in the interest of safe air travel, and he had succeeded only in denting their armor.

In late 1975 Mr. Ryther was interviewed, for the purposes of this book, in his home in McLean, Virginia, and asked to what extent the FAA had acted on the recommendations in his five-year-old report.

Ryther, a quiet-seeming man who looks to be in his mid-fifties, smiled for a moment and then said, "When I decided about a week ago to let you people interview me, I made a telephone call that same day to an Assistant Administrator at the Department of Transportation, a man I've known for years. I told him that you wanted to know what happened to my report, and I asked him if he would call his friends at the FAA and put the question to them, ask them if any of my original recommendations had been adopted or acted upon. That was a week ago, and I haven't heard a thing in response."

One of the Ryther report recommendations that he still feels very strongly about is the idea of an annual review of all FAA regulations to update whatever needs updating in the interest of safety. Ryther acknowledged that the vast majority of regulations would not need to be changed, that even though experts might find that "90 percent or perhaps even 98 percent were fine" the review "would permit you to zero

[44] *CLEARED FOR THE APPROACH*

in on those that needed change, on a regular basis, and not after some accident happened.

"So it was our proposal that the entire manual be examined annually. And that there be established a means and a procedure whereby those changes that needed to be made or where there were voids in new regulations promulgated, that they be done promptly."

What really bothers Philip Ryther to this day is that the sluggishness he uncovered five years ago is by no means gone:

"That sort of thing is related to the story of TWA 514. Because if that sort of review had been made, and honest-to-god objective decisions made concerning it, the thing that TWA had been appealing to the FAA to resolve would have been settled—we wouldn't have had one set of criteria in the hands of the controller and a different set of criteria in the hands of the pilot, and both of them acting in good faith confusing each other and resulting in the accident.

"There's just any number of other examples that could be cited where absolutely keeping up-to-date the regulations points right directly at crashes, or the avoiding of them."

The second point that still rankles Philip Ryther is that of the uncertified flight schools. To this day, and despite the concern among people like Ryther both within and without the FAA, a large number of schools teaching flying in this country do not have to meet any government standards. And what makes this so terribly dangerous is that many of the same people who teach, also rent and sell airplanes; thus they have a vested interest in seeing their students pass. It is not a desirable situation, and certainly not a safe one.

One of the most intriguing and frightening stories that Ryther told on the day he was interviewed had to do with what he viewed as the FAA's attempt to intimidate his publisher (Doubleday) when he was about to publish his book.

It was a wide-open secret in the first weeks of 1972 that Ryther's book was going to roast the FAA. The *Look* magazine article had said as much, and anyone who knew the man had to realize he would not be pulling any of his punches. Nonetheless, it was still a surprise when Evelyn Metzger, then Doubleday's Washington representative and now the head of her own publishing company (EPM Publications, Inc.) received a most pointed letter from a Mr. James A. Greenwood, the FAA's Director of Public Affairs.

Dated January 20, the letter read in part:

This will confirm our telephone conversation of last Thursday, 13 January 1972, regarding a manuscript which reportedly "exposes" Federal Aviation Administration deficiencies and challenges the integrity of FAA management. We understand the work is coauthored by Mr. Philip Ryther, a former FAA employee, and Mr. Stephen Aug, a reporter for *The Washington Star*. Naturally, we were disappointed that you feel certain material pertaining to the FAA, and purported to be factual, requires no bilateral verification. I'd think Doubleday would want to take precautions and insure that the text contains no actionable, libelous statements maligning public officials. . . .

The FAA spokesman went on to accuse Ryther of "self-serving concern for aviation safety," which he termed "classic examples of a deception calculated to achieve personal gain and aggrandisement." The writer offered to make available for Mrs. Metzger and Doubleday's inspection "the same records and briefings provided certain congressional committees that had expressed an interest in Mr. Ryther's charges."

The next to last paragraph read:

Seems to us that any reputable publisher of stature has an abiding responsibility to the general public. While the market for Mr. Ryther's vendetta appears marginal at best, we believe you might save Doubleday some embarrassing aggravation simply by double-checking a few facts. (I've yet to see a "hatchet job" of any kind founded on truth.) We trust you will reconsider our invitation to examine the record so that you might be better equipped to make editorial judgments."

Evelyn Metzger, who knew her way around both publishing and Washington, responded with a March 3, 1972, letter that, in contrast to the FAA missive, eschewed all bluffing.

Dear Mr. Greenwood:

In response to your letter of January 20, the contents of which took me completely by surprise, I have gone through all my correspondence and talked to a number of people to be able to verify the number of times that the Federal Aviation Administration was approached for, as you put it in your letter, "bilateral verification."

I would like to make one thing clear at the outset; Doubleday has taken precautions to insure that the text contains no actionable, libelous statements maligning public officials.

> The statement that "everyone's entitled to his opinion" does not in any way dilute Doubleday's fine reputation for high quality, objective non-fiction. Your impression that prior to publication Doubleday's policy is to research, document and authenticate a proposed work, is absolutely correct.
>
> As I have stated before, I am surprised that at this juncture, the Federal Aviation Administration has decided to allow its files to be reviewed when I have documentation, from sources including Mr. Ryther and Mr. Aug, that the F.A.A. denied access to these files previously.
>
> Specifically, I shall refer to your letter of March 9, 1971, in which you quote to Mr. Ryther pertinent sections of the Department of Transportation Regulations and the Public Information Act, as to the fact that in your determination the reports requested were submitted in confidence and accordingly should be withheld from public disclosure.
>
> There is also the fact that Mr. Aug had requested certain information from Mr. Archie Leauge of the F.A.A. and that similar sections of the above were quoted to him as a reason for denying him access to this information.
>
> The number of letters that were sent back and forth between the respective parties definitely indicates Doubleday's attempts, through its authors, to obtain the data that you are willing to allow Doubleday to see at this time.
>
> Now that you have changed your mind and are willing to release the information unsuccessfully sought in the past I would appreciate your getting in touch with me immediately to arrange a time that the material can be seen.
>
> Just to dispell any doubt which your letter may have created as to the depth Doubleday has gone to assure proper publication, the work has been subject to legal vetting [careful examination] on the part of the author's counsel and is at present being read by Doubleday's counsel.

That was the last the authors or their publisher heard from the FAA. What is incredible to me is that a government agency would attempt such strong-arm tactics *in writing*.

Ryther feels that Doubleday got the treatment from the government because he was making waves, and he bases his speculation on the fact that his last substantive White House contact was with Mr. John Dean.

A few years ago, I might have scoffed at such a suggestion. Today it seems quite plausible. Unfortunately, we are not likely to uncover the truth of the matter. The point remains, however, that a government agency tried to scare a major American publisher.

In view of Ryther's stinging but correct criticism of the two things he most disdained—the marginal air taxi operations and the uncertified flight schools—it is only fair to update the FAA's progress in these areas in the years since Ryther's book appeared.

As one who held an air taxi license during the late sixties, (when I leased my own planes for private charter flights) I can distinctly recall the FAA's sharp effort to tighten up the rules in order to bring all of us Part 135 certificate holders up to Part 121, the air carrier standards. Numerous changes have been required over the last five years, and one of the specific differences that I remember is that we, the air taxi operators, were required to substantially rewrite our operations manuals, the book that tells each new pilot how we run our particular air service. This extensive and costly project had the desired effect of discouraging marginal operators from attempting to go through the process, which they had to do in order to retain their right to fly for hire.

Various operators were affected by this move, and among those most profoundly affected were the so-called feeder air lines, small lines that operate on schedule between metropolitan areas and small communities, usually flying ten- to twenty-passenger twin engine aircraft. Although not subject to the regulations of the CAB, they were also not protected by exclusivity of routing. The feeder airlines are in a highly competitive situation, and the fellow who could operate most cheaply was the one most likely to make a profit at the end of the year—which meant that prior to the FAA rule-tightening there was an economic incentive to cut corners and a concomitant safety hazard.

Those feeders that were sufficiently organized—and capitalized—to survive the reformation very sharply upgraded their procedures, the currency of the training of their personnel (in some cases also the basic personnel qualifications), and the quality of their equipment. All of these changes heightened the safety quotient.

In addition to stiffening the rules, the FAA began making a series of spot-checks on air taxi operators, which in many ways paralleled those made on the air lines. At the same time, the Agency permitted the feeder airlines to do a good deal of its own checking—through FAA-designated check pilots employed by the feeder airlines. Again, this was much like the process used to check the pilots and crew on the major airlines.

These changes helped to alleviate the problems of the gypsy charter lines, but the other problem cited by Mr. Ryther—that of the uncertified flight school—remains.

Although I think there has been some improvement in the operating standards of these schools in the past five years, they are still not subject to the kinds of regulations that a certified school must live with. The only encouragement given by the FAA to seek certificate status is that in some cases a shorter training time may suffice. Certified flight schools are allowed in certain circumstances to graduate students with a minimum of thirty-five hours' flight time. Still, this is not necessarily a clear-cut advantage, because most students require about fifty hours before the school (certified or not) will submit them for flight examination.

In the main, flight schools have become cautious, because they are sufficiently jealous of their certificates and usually resist the pressures of a student who wants to graduate early, even if he or she is still "marginal" in one or more areas.

The argument has been made for a long time that the FAA doesn't want to impose its standards on each and every flight school because there are many small schools in remote areas that operate on a part-time basis, and that to force them to meet the standards of schools in larger areas would thereby impede the growth of aviation, a "basic American right."

It seems to me that when the FAA takes this rather uncharacteristic stand *against* regulation, it fails to balance its priorities. Certainly the growth of aviation has to be allowed and promoted, as it has always been, but in modern times the skies are sufficiently crowded and users of the system are such a potential hazard to one another, that I feel the FAA should deal with the existing flight schools in exactly the same way it dealt with the existing air taxi operators five years ago—either come up to speed or get out of the business.

Back in his FAA days, Philip Ryther had a favorite bit of personal philosophy copied and framed. He hung it on the wall of his office at the FAA. Taken from a column in *Newsweek* magazine, it read:

> A great number of otherwise well-qualified men never seem to rise above a certain level. They seem to have a deathly

fear of ever being wrong. They never want to make a decision because they never want to make a mistake. If it is your goal in life to be known, to your associates as the man who never made a mistake, then be prepared to be known also as the man who never really accomplished very much. Today, make a few decisions on your own. They may be right or they may be wrong. But you made them. And tomorrow your boss may also give you a raise.

This brief bit of individualistic philosophy is entitled, "The Man Who Never Made a Mistake."

Not too long after he had hung it in a prominent place in his FAA office, Ryther was told by his superior (his boss) to take it down—it was "not suitable for a government office."

* * *

If the fate of the "Ryther Report" reads like an exercise in frustration, consider the case of the House subcommittee report from which I quoted at the beginning of this chapter.

Even before the crash of TWA 514, the subcommittee was convinced of the pressing need for certain internal reforms at the FAA.

It had already concluded that the administration's handling of the cargo-door problem with the DC-10's was improper, or at least sorely deficient in view of its mandate to protect the public. And the subcommittee had drafted hard recommendations in regard to the issuance of Airworthiness Directives, the handling of hazardous materials, the FAA's response to NTSB recommendations, the problem of granting too much time for the industry to comply with safety directives, and the FAA's "dual responsibility for passenger safety and for the economic well-being of the air line industry."

The tragic news of the 514 crash brought a renewed effort on the part of the staff, the subcommittee, and Chairman Staggers for swift action on the question of whether or not to require the installation of Ground Proximity Warning Systems on all major commercial airliners. Statements were made on the floor of the House of Representatives by both Staggers and committee member Jake Pickle, a Texas Democrat.

Other voices were also raised, and within days, the FAA took action. The unusually swift response was made by Administrator Alexander Butterfield (the man who had stunned the Watergate Investigating Committee, and ultimately the nation as well, by revealing the secret taping system within Richard Nixon's White House) who flew out to Redmond, Washington, for a personal demonstration of the Sundstrand system.

Butterfield came away convinced, and before leaving Washington State he cabled Staggers' subcommittee to inform them that he was ready to make the GPWS mandatory equipment. He did so on December 18, 1974, in a mandatory rule which required all major commercial airliners in the United States to have a Ground Proximity Warning System installed by December 1, 1975.

This was most heartening to the subcommittee, for it appeared that finally there was someone at the head of FAA who was not afraid to make a move that the industry might not like.

They were encouraged by the initial press reaction to the report and by a letter to the subcommittee from Claude S. Brinegar, the Secretary of Transportation. Less than a week after the official release date of the Report (December 28, 1974) the Secretary wrote to Chairman Staggers. The main paragraph read:

> Please let me assure you that the serious issues about the FAA raised in this report are receiving my personal attention. As a first step, I have convened a meeting of FAA top management, including the regional directors on January 15, 1975. Following that meeting I will appoint a special task force to study the various issues in depth in order to have firm recommendations ready to present to my successor as early as possible.

(Brinegar's mention of his leaving was no secret, he had announced that he would be gone by February 1, but it was an interesting note in light of the rumors that Alexander Butterfield's days were numbered.)

For once, the Washington machine seemed to be moving in high gear. On the day after the secretary wrote to inform the subcommittee that he was going to appoint a special task force, FAA head Butterfield named a panel of his own. And it was a beaut. Unlike so many other panels and task forces, this one had three members who really knew their field of study. Headed by Richard Shoup, a former congressman who had been concerned with air safety for years, the panel included David Thomas, formerly a deputy administrator of the FAA, and Paul Sonderlind, once the chief pilot of Northwest Airlines.

Butterfield told the trio to make recommendations to correct "any deficiencies in the FAA from the standpoint of safety." No sooner had they begun, however, when they learned that Brinegar was not pleased with the scope of their assignment. As Shoup later told the UPI, "After we had started work, the Secretary of Transportation limited the scope of our investigation to making sure the dates and times in the congressional report were correct. He also said he would name his own panel to investigate the things we were looking into. We felt we could not do anything to contribute to the safety of the flying public under those limitations and that we would just waste the taxpayers' money because we were so hampered."

The Butterfield panel, not yet two weeks old, resigned in protest.

That left the issue of the subcommittee's recommendations up to the DOT task force. The group was named on January 28, 1975, one month to the day after the release of the subcommittee report. By this time Butterfield had resigned, and the official press release from the office of the Secretary of Transportation identified the task members as follows:

CHAIRMAN: Lt. General Benjamin O. Davis—Assistant Secretary for Environment, Safety and Consumer Affairs. Former Chief of Staff for Korea and Far East.

EXECUTIVE SECRETARY: James Dow, Acting Administrator, Federal Aviation Administration.

OTHER MEMBERS: Warren G. Bennis—President of the University of Cincinnati. Author on the subject of organizational and management issues.

Michael Collins—Director, Air & Space Museum, Smithsonian Institution, Washington, D.C. Former astronaut and Assistant Secretary of State.

Edgar M. Cortright—Director, NASA Langley Research Center, Hampton, Virginia.

> Willis M. Hawkins—Aviation Consultant. Retired Senior Vice President for R&D at Lockheed. Former Assistant Secretary of the Army for R&D.
>
> George A. Warde—Aviation Consultant. Former President of American Airlines.
>
> Louis B. Young—Aviation Consultant. Retired Vice President of Bendix Corporation. Former Chairman of the General Aviation Manufacturers' Association.

I think it fair to say that, in comparison with the FAA panel, the DOT task force was not exactly heavy on the side of air safety investigation expertise.

On his way out the door, Brinegar had charged the task force with studying the House subcommittee's recommendations in order to come up with its own conclusions and recommendations. On April 30, 1975, the task force publicly submitted its report.

Of course no one expected the DOT—the department that contains the FAA and in theory oversees it—to let the agency have it right between the eyes, nonetheless there was marked disappointment in many sectors of the aviation world. For one thing, the task force's 19 recommendations were nowhere near as specific and demanding as those contained in the subcommittee's report, and for another, the tone was one of outright cordiality.

The task force began by praising the FAA. "Over the years, FAA has achieved a history of remarkable accomplishment in its dual role of both fostering and regulating aviation. To grasp the dimensions of the FAA's accomplishments, it is only necessary to cite a few examples . . ." The report ended on much the same note.

Not too surprisingly, the report recommended that the FAA—which accounts for 78 percent of the employees of the DOT—be left within the umbrella of the Department of Transportation, rather than being made into a separate governmental entity. Nor was it a shock to read that the task force saw no need to grant independence to the National Transportation Safety Board. Also not very surprising was the report's implied suggestion (found between many of the lines) that what the FAA *really* needed was more personnel, equipment, and money. The report was a classic bureaucratic response to outside criticism.

Secretary Coleman, the new head of DOT, told the audience at the press conference called to officially release the report that "I basically concur with its findings..."

Then he said:

> General Davis will comment in a moment on the specific recommendations. I just want to say two things with respect to the report and its conclusions:
>
> First—while I did not instigate this review or appoint the members of the Task Force, I have communicated to the members the importance of producing an independent and objective assessment of the FAA's approach to air safety.
>
> I have stressed before, and do so again today, that there can be no compromise with safety. If there are any doubts concerning the meaning of FAA rules, then let us go the second mile to assure clarity and understanding. If too much time is being spent in the rulemaking process, or in getting direction to the field, then let's do what's necessary to shorten the time.
>
> Second—I came to this job aware of rumored discontent between the Office of the Secretary and the FAA. If frictions have flared in the past, I'm sure they can be avoided in the future. The Task Force has found no basis or merit for a separation of FAA and DOT. We are of necessity linked by a common desire to move together toward the balanced, efficient, integrated national transportation system we all desire for our country. As you know, there are 55,000 employees in the FAA. It makes major policy decisions, determines certification of aircraft, controls the American airways and is the prime mover on aviation safety matters. It is both a policy developer and regulator. It thus must have a spokesman in the Cabinet and before the Congress. Obviously, this is the mission of the Secretary of Transportation. I am confident that my office, and that of the Administrator can function without conflict, as we move concurrently toward the goal of better transportation and increased air safety.

So much for the question of an independent FAA; and so much for the congressional criticism of the agency. But then, when a task force ostensibly charged with investigating complaints against a government body, chooses as its executive secretary the acting head of that body, what should anyone expect?

There are two interesting and potentially disturbing footnotes to the saga of the House subcommittee report on air safety.

The first has to do with the fact that shortly before the report was finished, the subcommittee learned that under a new reorganization plan within the House of Representatives, it was to lose its jurisdiction over air safety to the aviation subcommittee of the House Committee on Public Works and Transportation.

Without going into the Byzantine maneuverings that characterize any attempt to change the way Congress does its business, it is sufficient to say that by moving air safety to the purview of another committee, the powers-that-were all but assured a less intensive follow-up of the subcommittee's recommendations. It is a simple fact of life that those who did not do the study or write the report will not be as eager to push the issues raised.

In fairness to the new committee and its staff, it has held several get-acquainted hearings, and it has also hired at least two people with some experience in aviation. But it will be a long time before its experience matches that of the staff that turned out the air safety report.

No one seems to feel that there was a dark plot behind the move to switch the jurisdiction, but the fact remains that it was, in essence, a break for the FAA. And it should surprise no one that most of the staff members who did the study on air safety have left the Congress and now work elsewhere. Their expertise on certain matters of air safety is diminishing, their value decreasing.

The second point is more intriguing. It concerns the FAA rule requiring the installation of the Ground Proximity Warning System on all major airliners by December 1, 1975.

As far as the average airline passenger is concerned, that requirement—if he or she even read about it—probably sounded as if the FAA had finally paid attention to one of the prime recommendations of the House report.

Unfortunately that was not quite the case. At the time the report came out there was only one ground prox device on the market and ready for purchase—the Sundstrand product. Their box contained five functions, or modes, designed to cover the full range of potential accidents. The modes covered excessive sink rate; excessive terrain closure rate (the 514 situation); negative climb after takeoff; inadvertent proximity to ground without gear or flaps; and inadvertent "duck-under" the glide slope, (which simply means slipping below that "electronic road" which an airplane flying on instruments follows from the sky to the earth).

When the FAA ordered all the major airlines to have GPWS systems installed and functioning by December 1, 1975, it required only the first *four* modes.

At the time it did so, the FAA said that the mode five technology was not sufficiently advanced. This was, as one knowledgeable observer noted, "ridiculous. Sundstrand had already been selling five mode devices to Boeing and Pan Am well before the requirement, and nobody was complaining."

Apparently some captains habitually deviate from the glide slope when landing for reasons that do not jeopardize passenger safety (for example, when landing at high altitude fields in the Far West or in certain foreign countries). Shortly after the ground prox device had been installed in the planes of some of the early premandatory customers, the pilots found that in these situations the device would go off. Pan Am found this happening, but the occurrences were so rare, and so easily solved by the pilots' making a simple adjustment, that it was not considered a problem.

According to several sources, this is what the FAA was concerned with when it refused to require the installation of the fifth mode.

Exactly why the FAA took this position, however, is not clear. It is clear that the traditional suppliers of avionic equipment feared that the small Sundstrand company was being handed a government monopoly on a silver platter. (In fact, that's what it would have been in the beginning, but it would have been justified on the basis of expediency, for it would have been a *temporary* monopoly in the interest of air safety.)

Some heavy lobbying took place. Even before the rule went into effect, one of the large electronics manufacturers showed up at the offices of the House subcommittee with what one staffer called "a Chinese copy of the Sundstrand device."

Rumors began to circulate that Sundstrand had actually pressured the House into holding its hearings as a way of peddling its product by government fiat (the eventual FAA requirement). Eventually the rumors became allegations which Sundstrand quickly and openly denied.

What had happened was that the small company had pioneered and extensively tested the device, and it had supplied the House subcommittee with the results of its research indicating the need for such a device. (It found that terrain crashes accounted for 56 percent of all aircraft accidents, and that these occurred while the plane was apparently under control.) And it did send its people to Washington to testify at the hearings.

This meant that when the mandatory rule was passed, Sundstrand was in a very enviable position, as far as the other manufacturers of avionics were concerned. Indeed, it was not only number one, it was the only one.

In the March 17, 1975, issue of *Business Week*, the New Products column was devoted to the flap between Sundstrand and the large companies. Entitled "A Lesson in How Not to Sell to the Airlines," its first few paragraphs summed up the issues:

The carriers don't like the way Sundstrand put FAA pressure on them.

In December, Sundstrand Data Control, Inc. was flying high. After investing $6-million and six years to prove out a crash warning system for aircraft, it had sold some 500 of the systems to Pan American World Airways and to 37 foreign carriers. Boeing Co. had made it standard equipment on all its new commercial jets. Best of all, the Federal Aviation Administration had just ruled that such a system must be installed on all U.S. commercial aircraft by Dec. 1, 1975. With that short lead time and no competition in sight, the Redmond (Wash.) subsidiary of Sundstrand Corp. hoped that it had a $20-million market in the avionics devices all to itself.

That hope has now crashed. The carriers indeed have been rushing to buy the so-called ground proximity warning system (GPWS). But of the roughly 1,000 units contracted for since the FAA fiat, almost all in the past three weeks, Sundstrand has sold fewer than 300, and many small orders account for the total.

The bulk of the business so far is going instead to Bendix Corporation's Avionics Div., Rockwell International's Collins Radio Div., and Litton Industries' Aero Products Div. "This is frustrating for us," acknowledges William R. Kopp, Sundstrand Data's general manager. His company's GPWS is the only one that has been certified by the FAA and the only one that is in production. Even more telling, Sundstrand's device has racked up more than 1-million hours of flight time. "We are fighting paper designs," Kopp wails . . ."

FEAR OF FLYING [57]

What had happened is that by withholding certification of the fifth mode, the FAA (for whatever its reasons) held the door open for the larger manufacturers to come in with their own devices. This effectively neutralized the small company's lead.

Sundstrand has been fighting ever since to hold on to its fair share of the market, despite some rather curious happenings.

In May the FAA did a flip-flop and announced that it would now require the GPWS to include mode five, the warning against glide slope deviation. Apparently the technology gap had been solved in little over six months' time.

One might surmise that this had to help Sundstrand, simply because it was the only company with a finished box, certified by the FAA. One would be wrong. The FAA provided the equalizer on November 3, 1975, by announcing that it was extending the deadline for compliance under certain circumstances from December 1, 1975, to June 1, 1976.

The official FAA press release read:

> Due to equipment shortages, the Federal Aviation Administration of the Department of Transportation announced today that it will approve extensions on an individual basis for up to six months of the December 1 compliance date for installation of ground proximity warning systems (GPWS) in large turbine-powered airplanes operated by the airlines, air travel clubs, and air taxi operators.
>
> However, extensions will be granted only to those operators who demonstrate to the FAA that they cannot comply with the original GPWS installation date due to circumstances beyond their control. This showing must be made by December 1 and the operator also must submit a schedule for compliance indicating that the system will be installed at the earliest practicable date.
>
> FAA's action is based on a petition from The Air Transport Association of America, which stated that delays in certification programs together with delays in equipment delivery by some manufacturers have caused difficulties for some ATA member airlines in meeting the December 1 deadline. ATA pointed out that responses to questionnaires sent to all member airlines indicated that 78.5% of the total airline fleet will have the GPWS installed by December 1. The airlines operate approximately 2,400 aircraft.
>
> ATA further stated that a secondary benefit would accrue

to operators receiving extensions by permitting concurrent installation of the glide slope deviation alert system (required by June 1, 1976, under another FAA rule) and thereby, "lessen the need for removal of aircraft from service, potentially reduce the need for dual certification testing, and simplify crew training procedures."

No matter how one views the position taken—and then changed—by the FAA, it's hard to deny that the FAA helped the large companies that traditionally supply the major airlines and hurt the people at Sundstrand.

This action, of course, is defensible on the basis of the need for free competition in the open market. And FAA extensions are not rare; they are made all the time for perfectly legitimate reasons.

But if a plane had gone into a mountain between December 1, 1975, and June 1 of the next year, and that plane was not equipped with a ground proximity warning system because of "manufacturer's delay," the FAA would have come in for some heavy criticism.

And, of central importance, some more people would have been killed.

There is a potential for cruel irony in the situation: Had the FAA not been so mindful of free enterprise, Sundstrand could have supplied all 2,400 aircraft in time to make the original December 1 deadline.

What is so chilling is that we have been down this road before, specifically with the cargo-door problem on the DC-10's. Remember the FAA and McDonell-Douglas knew about the potential danger after the incident in Detroit, and yet they agreed to a "Band-Aid fix."

In November 1975 one of the former House subcommittee staffers said, "It scares the hell out of me because it's beginning to sound like the DC-10 situation all over again."

Let us pray that the man is an alarmist.

* * *

I must make it clear that my less-than-enthusiastic attitude toward the FAA is not an all-encompassing criticism. The great majority of the employees are sincere and dedicated to the cause of air safety—after all, many of them do far more flying than the average citizen—but there are too many people in positions of power who are too quick to take the side of the industry, and in one way or another forget what should be their priorities.

For example, one of our sources told us that when he accompanied a man from the Sundstrand corporation to a meeting at the FAA one afternoon, he witnessed a most unsettling incident.

The man from Sundstrand made the passing comment that in time his company's device would not be the only one certified by the FAA and on the market, whereupon the government official stood up and in a forceful tone said, "You can be damn sure of *that!*"

And then there was the man from the House subcommittee who called a minor official in the FAA to get some information about the administration's position in regard to the GPWS and was told, "You people just don't understand what you're doing. Something like this could adversely affect the international balance of payments." The congressional investigator told us, "I could give less of a damn about the balance of payments. I expected to hear the man say he was primarily concerned with saving human lives!"

And finally, in a different context, Philip Ryther tells of the time he was inspecting an airport in the far Northeast in the company of an FAA regional director. The government had just decided to close the federally owned field, and in fact the major airlines had signs all over the place announcing the dates on which they would discontinue their service. Yet, as Ryther saw with his own eyes, workmen were painting the fences alongside the terminal, while out on the main runway of the soon-to-be-closed field, another crew was installing a new landing system.

He asked the regional director what was going on, and the man informed him that the FAA had an obligation to the people of the area, which was suffering from chronic low employment. As Ryther said, "I admired the man's compassion, but I really wondered if that was a proper role for the FAA to undertake."

The FAA suffers, perhaps more so than any other federal agency or administration, from an inherent tendency toward foot-dragging. There is an organizational lethargy about the place that must be eliminated if the needed changes are to take place.

When and if the time comes to rewrite the basic legislation behind the FAA, I hope Congress will give serious thought to the idea of eliminating the FAA's dual role and charge it only with fostering air safety.

The aviation industry is big enough to take care of itself now. It doesn't need the government to promote it.

Let the FAA worry only about safety.

4

The Human Factor

I had a license to fly military jet aircraft by the time I was twenty years old, and although I would not trade that very special training for anything in the world, there were some slight difficulties the first time I tried to fly a light civilian-type aircraft.

The Navy had a rule that you were not allowed to fly any civilian airplanes while you were in flight training. Even though some of the guys already had their private licenses, the Navy felt that for obvious reasons it simply didn't want any of its flight trainees ending up in the newspapers for having cracked up a light airplane. Not only wasn't civilan flying allowed, but you could be kicked out of the program for it.

Of course some of the men in my class already had civilian ratings when they came in, and when they took their familiarization flight they embarrassed the instructor on the first day. Most of us, however, had learned to fly in the military, so we abided by the rule; but we looked forward to the day we could also fly the smaller planes.

The minute you were given your wings and designated a naval aviator, you could take a very short written test for a commercial license, and later on you were also given an instrument rating. Everyone took the test, and I was no exception.

My first flight was not exactly auspicious. And the second was worst.

I went in to rent an airplane and the man in charge said, "We have to check you out."

[62] CLEARED FOR THE APPROACH

I protested that I was a military pilot, and that I'd just gone through the most rigorous flight training in the world. He said, "That may well be, but our insurance policy says we still have to check you out." And it's a good thing he did.

The first plane they took me up in caught fire. It was an Aeronca Champion, which he chose because it was a tandem trainer and very similar to what I'd learned on. But we got up in the air and it started smoking, so the instructor grabbed the controls and back down we went.

Then they gave me a Cessna 120, a pretty little silver bird. I had been used to flying heavy military airplanes with *lots* of power, very tricky to get on and off the ground. At first it was Air Force AT6 Texans—a fore-and-aft trainer with a green house for a canopy, a big radial engine, and terribly narrow landing gear. The plane was made by North American Aviation, and they made tons of them. In fact, in all the old war movies whenever you saw a Japanese Zero, it was actually one of these planes. Not only did they become a very popular racing plane with a class of their own, but certain Latin and South American countries are still using them as military fighter planes.

We flew those planes all through basic training, and then we switched to World War II fighters—Grumman Hellcats and Bearcats—which were really a repeat of what we'd been flying before, but suddenly instead of 900 horsepower we had over 2,000.

After that kind of experience, the tiny Cessna looked like a toy. And I thought it would probably all but fly itself. I was quite wrong. In the stuff I'd flown, you pulled the power on and you came down like a stone. But the Cessna had little bitty engines and great big thick wings, and it just fluttered over the mile-long runway. On my first approach I missed the field entirely.

As I said, it was not an auspicious beginning, but after that things got better. Probably because they *had* to.

* * *

Before getting into the training and background of the pilots most readers come in contact with or happen to know, I want to spend a bit more time describing the kind of military training I received, because it points up several factors generic to all pilots and several areas in which they differ.

I've said many times that if I ever started a training school for trial lawyers, I would insist that each student also learn how to fly. I say that because a good flight school, and pilot training in general, inculcates the idea that "I am functioning on my own and I can't get any help; therefore if I lose my confidence I am in very bad hands—my own. My decisional ability diminishes, as does the smoothness with which I handle the controls and make the airplane do what I want it to do. Ultimately, the whole cutting line goes."

Confidence in your own ability, based on training and native skills, is essential to a pilot. It *should* also be essential to a trial lawyer, but there are too many ill-served clients who can attest to the fact that it isn't. And, ironically, we license pilots but we do not license trial lawyers.

If I check somebody out in an airplane, when it comes time to turn him or her loose with it there's only one question in my mind, "Does this pilot control the airplane? Does he master it, control it, make it do what he wants? Or does he follow it around the sky?"

When Dave Savage (my chief pilot) or I take somebody up to check them out in one of my airplanes, we don't just fly around the field. We take the airplane and suddenly pull it into a pitch-up, or kick a rudder to throw it into a spin, and say, "You've got it." Now, does he get the airplane back, or do we have to take it away from him and restore normal flight?

The essential difference between military and civilian training is tantamount to the difference between war and peace: in the military, you are training a person to function under extreme emergency conditions, often involving imminent danger to the individual's life and limb. Therefore, the military compels the student pilot to fly in conditions that are far more exacting than those usually presented to the civilian trainee.

First of all, the military makes him take chances all over the place. They deliberately push a pilot to the outer edges of the strain of flying. They make him fly right at the "red line." They make him, for example, shoot gunnery runs at a target, and if he doesn't break off at exactly the right instant, he's going to crash. They'll put him up in the worst weather in the world (when "even the birds are walking") just to see if he can take it. All of which, of course, makes sense in training a military machine, where the individual in the plane is relatively unimportant in comparison with the overall success of the mission or objective.

Even though I was trained during peacetime (the early 1950s) we were put through the wringer on a very thorough basis. There was no pressure to crank out the pilots, like there had been during World War II, and even then the shortest training period was six to nine months. Our program was eighteen months long and exhausting.

By the time you came out you were an all-weather pilot, capable of landing on a carrier in several different kinds of airplanes, capable of making straight-in-runs, shooting at an air-to-air target, dive-bombing, rocketry, and formation flying. If a civilian flight school tried to put together such a program, it would be extremely risky and exorbitantly expensive.

All of which is why whenever I really get in the soup in one of my own planes, I am damn glad to have had the military training and to know that Dave Savage seated right next to me had that same background. This is not to say that all pilots should have flown in the service. There are any number of superb pilots, some of them flying as captains for the major airlines, whose background and training is entirely civilian—I think particularly of Dick Bedell, who is equally at home in a Rockwell Commander and an Enstrom helicopter, and what's more can tear each one down and put it together again. But red-line training has its own special rewards, and I will never regret having received it.

One final note on that training—even at the risk of belaboring the point—once you got your wings you were by no means done with your training. An ocean of training and retraining goes on all the time in the military. Once you are designated sharp, you are constantly retested and retrained to make sure that you stay sharp. The byword is currency, currency, currency. And that is the greatest point of similarity between the military pilot and the man who flies for a major airline, but I will be going into that point in much greater detail later on in this chapter.

As in any other line of work, there are pilots and there are pilots. To begin with, one should realize that almost anyone can learn to fly (see the chapter entitled, "General Aviation"). It does not take an extraordinary degree of manual skill to fly an airplane. And neither age nor physical defect (up to a point) is necessarily a barrier. But there are

FEAR OF FLYING

vast differences in skill and attitude among the 733,728 souls who hold a license to fly in the United States.

There was a distinction in the military between those who wanted to be fighter pilots, per se, and those who took what might be called the longer view. The former always opted for advanced training in jets (myself among them) while the latter chose advanced training in multi-engine aircraft, which we fighter pilots sneeringly referred to as multi-engined boats. The navy had trouble while I was in training in getting pilots to go into the multi-engine program, but most of those who did go in already had their minds made up to be airline pilots one day. And they were smart enough to realize that the airlines would be a lot more impressed with that experience than they would be with several thousand hours in a small fighter plane. Essentially, and I'm by no means putting the knock on them, the men who chose this kind of training were the cautious plan-ahead types. And that's not a bad type to have up front when you're flying commercial and there's bad weather brewing over Chicago forty minutes ahead of you.

Before turning to the training of the airline pilots, let's take a quick look at a very special type of pilot—the flying superstar. Within any category there are different levels of skill, even though the group as a whole is highly qualified. (Fran Tarkenton makes the rollout run look easy; in his playing days with the Redskins, Sonny Jurgensen was known as a premier passer, but on the rare occasions when he had to run, the entire stadium broke out in fearful laughter.)

I like to think of myself as a pretty damn good pilot (all pilots do) even if I have to admit that my "hot pilot" days are in the past. But I have flown with some people who make me feel as if I were still wobbling over the field in that silver Cessna at Corpus Christi, Texas.

Two such pilots are Clay Lacy and Bob Hoover; another is Mike Meger, in a helicopter.

Lacy is a United Airlines captain, but he is also an Enstrom dealer (in partnership with Hoover) in California. I once had to charter Lacy's Lear Jet, to get from Los Angeles to Las Vegas in a big hurry. He cut the normal rate of $660 an hour down to $200 (which barely paid for gas and fees) because I could fly copilot and because he's a nice guy.

I'd heard a great deal about his skill as a driver of airplanes, so I wanted to put on my best performance. When I greased the plane onto the runway in Vegas, I thought I'd acquitted myself quite nicely. Clay, a gentleman, didn't say anything one way or the other.

On the way back, however, he simply showed me what flying a Lear Jet is all about—at his level.

The cockpit in a Lear has never been called roomy, and Lacy is a great big guy. He had to jam himself in the seat. Then he took off and flew the airplane, and I swear he didn't even seem to be touching the controls. He was sitting there, all knees and elbows, and he'd put a little finger on this or a thumb on that and the Lear reacted as if it were flying itself. His performance was simply gorgeous. But that's not a seasoned pilot; that's a superstar.

Lacy is the president of the Professional Race Pilot's Association, but he is probably not as well-known as his friend and partner, Bob Hoover. Anyone who has ever seen an air show anywhere around the country has undoubtedly seen Hoover perform. One of his regular stunts is to cut a ribbon held ten feet off the ground across two poles with the tail of his P-51 "antique" fighter plane. (Note: in order to do that you must be flying upside down.) Another specialty is to put his plane through an exacting barrel roll, which means turning a complete circle. But Hoover adds a little something to the trick—while he's doing it, he takes a pitcher of water and a glass and he fills the glass while rolling over. I know it sounds like a cliché, but he does not spill a drop. I can't help the truth.

Mike Meger in a helicopter simply has to be seen to be believed. He shuts the engine down and drops the bird out of the sky from hundreds of feet, only to catch it at the last possible moment. He zooms over the field at fifty feet doing intricate maneuvers *backwards*. And his trick of bouncing first on one landing skid and then another makes the helicopter look as if it were dancing. Like Hoover, Meger is frequently seen at the major air shows. Recently both of them performed at the big one, Reading, Pennsylvania.

People like this are the superstars. All three of them are great big, confident, skilled, delightful human beings. It's almost as much fun to sit around and have a drink with them as it is to watch them fly.

But these pilots are the crème de la crème. The rest

FEAR OF FLYING

of the flying world is divided into two categories: the seasoned and everyone else.

There is a broad gap between someone who has a license as a private pilot and a truly seasoned pilot. Nor is a hot pilot necessarily a seasoned pilot. There are pilots who fly balls out all the time—moving airplanes through tricks even though the plane is not certified for aerobatics or taking a Lear Jet and rolling it eight times (which it does beautifully, but which is illegal as hell).

The seasoned pilot is one who hits a plateau. He stops being a hot pilot and wisely begins to believe in and depend on maxims like, "Don't get behind the airplane," and "You control the airplane," and even "There are old pilots and there are bold pilots, but there are no old, bold pilots" (which happens to be a bunch of horseshit, incidentally, as Bob Hoover proves every time he gets in an airplane at an air show).

To go from "licensed and adequate" to "seasoned" requires a transition wherein a pilot sheds his derring-do and replaces it with cautious confidence (or confident caution, whichever order you prefer).

This is not to say that he doesn't get tense. A good pilot can be tense and still be in complete control of the airplane. It may sound a bit grandiose to say, along with Hemingway, that a seasoned pilot faced with an emergency reacts with "grace under pressure," but that is just what he does. And that's the kind of person you want to have up front when and if there's trouble.

This kind of pilot—and it should be evident at this point that I am talking mainly about the men who fly for the major airlines and the large corporations—is comfortable in difficult situations. The common denominator of the people who fly comfortably and well—a group that includes virtually all of the airline pilots and the good corporate business pilots—is that they develop the same kind of confidence that I feel in the courtroom.

Some lawyers (some trial lawyers) are so uptight in the courtroom that they develop heart problems in their forties, whereas I'm completely at home in a courtroom. That's where I live, and I only tense up when it is appropriate to whatever is unfolding in the legal action, like a hostile witness.

When trouble occurs in the airplane, the seasoned pilot can

deal with it. It would perhaps be going too far to say that he actually enjoys an emergency—no pilot does—but there is a certain pleasure in being tested on hard-earned skills.

The seasoned pilot also develops another attitude, one that has saved countless lives, and that is a healthy resignation (not quite fatalistic, but tending in that direction) that everyone is going to go sooner or later. Please realize that I am talking about those rare, extreme emergencies. At such moments, the seasoned pilot says to himself, "Well, I chose this profession, and if my number is up, then it's up. The world is not going to stop turning without me. Now, let's face this situation head-on and see if we can beat it."

This is diametrically opposed to the other type of pilot who in the same straits says, "Oh, my God, my God, I might die. *I* can't die. What a horrible thing it would be if I died. How terrible for everybody, especially me!" and this really colors his thinking and causes him to panic, which often ensures his doom.

Panic doesn't solve anything. It's an involuntary response, like jealousy. You have very little control over it and it can make you act like a fool. Obviously it is far better to remain calm, cool, and collected—though I'll admit that there are times when to remain calm, cool, and collected one would have to be a pure psychopath.

The reaction of the seasoned pilot when faced with an emergency caused by weather or a defect in manufacturing or maintenance is that there is no point in getting mad or afraid. In such cases you are a victim, and you have to do whatever you can to avoid being a fatal victim. (A holdup victim who panics and grabs for the pistol, rather than quietly and quickly turning over his valuables, is vastly increasing his chances of losing more than his material goods.)

There are times when a good trial lawyer knows he's got to be just as firm with his client—sometimes for the client's good, sometimes for the good of society—as he has to be with the prosecution. And if he panics when things or people take an unexpected turn for the worse, he's not exactly a seasoned trial lawyer. Once again the analogy between trial work and flying holds true.

Most pilots who fly for the airlines today are either ex-military or very experienced commercial pilots who went through civilian training but built up enough time and a sufficient amount of different experience that they were judged acceptable for training by the airlines. And by background, I mean a substantial background, not just some pilot used to boring holes in the sky on a Sunday afternoon.

Not too many years back, the flight engineer position was a career job. But nowadays it is often the first step up the line toward captain; thus many airlines have in effect a third seat. Although the flight engineer doesn't really have any access to the controls, he is allowed to speak and to run the checklist. The pilot and copilot are the ones who physically put their hands on the controls—knobs, levers, buttons, switches, throttles, whatever. These days, flight engineers are more likely than not to be career pilots waiting their turn. And doing an important job in the process.

There was a time, not too long ago, when almost all the airline pilots were former military types, some of them closer in spirit to the seat-of-the-pants days than to the computer era. But those days are gone, and most of those pilots are in retirement, or close to it. These days, if you want to work for the airlines you have to make your choice early: do it while you're young, or you won't be able to get in.

It's like military flying. You get too old; you can be the best pilot in the world, but you won't get hired by the airlines because who needs a 52-year-old copilot? And what good pilot at that age could hack that status? So the airline captain of today is a highly skilled, extremely well-qualified career professional. Forget about that image of the dashing young man in the brown leather jacket and the white scarf trailing in the wind. Today's major airline captain is far more likely to be your slightly paunchy neighbor (provided you live in a pretty good neighborhood) with the clear-eyed gaze who always seems to be working on his lawn while you're inside the house plunked down in front of the boob tube.

Instead of generalizing, or creating a "composite captain" based on the statistics provided by the Airline Transport Association, let's look at one man who flies for a major airline, a man who has probably carried a good number of readers at least part way across the country, especially if you fly from one coast to the other.

* * *

"I took my first ride in an airplane when I was six years old.

"I grew up in Jamestown, New York, not far from Buffalo, and there was a little town nearby, maybe 300 to 400 people, called Stockton. And every summer they held the Stockton Town Picnic. My father used to take me; he was a car dealer, and he used to demonstrate cars there, sold a lot of them to the farmers, as a matter of fact.

"The Stockton Town Picnic had all sorts of events—a horseshoe pitching contest, a ball game, and a whole lot of food. It was held in a big open field just outside of town.

"One year my father took me and a friend of mine. And a guy came flying in, flying a J2 Piper Cub with no tail wheel on it, just tail skids, and he asked us if we wanted to go for a ride. It was like that *Waldo Pepper* movie, exactly like that. Well, of course he didn't have to ask us twice.

"We got in the back seat, both of us, and we took off on this old field. And that was it. From then on I just lived airplanes.

"When I was old enough, my mother would wake up on Saturday mornings and say, 'Where's Don?' But I'd be out of the house already, down at the airport in Jamestown. I got there in time to help open the hangar doors and to help push the airplanes out, and in exchange for that they'd give me a ride sometimes. A lot of kids did that in those days. So I did whatever I could around the airport to get rides, and I guess I just went on from there. . . ."

"Captain Don" is a senior captain for one of the nation's largest airlines. (Unfortunately, the airline in question would not allow us to use the captain's real name—nor even that of the airline itself. For some reason, even good publicity is suspect.) The interest in flying that began with the Piper Cub ride at the town picnic turned into a career. His route to the left seat, however, was not exactly a straight line.

He quit high school after his junior year to enlist in the Navy. The war was on and he wanted to be a part of the fighting. But he didn't learn to fly in the Navy, at least not officially, even though he had a good many hours in the air by that time. He served as a flight crew member, and when he got out he earned a private pilot's license (thanks to the GI Bill) and then a commercial ticket. (He also finished his schooling and received his high school degree.)

But he wanted more, and the GI Bill would pay for only so much. So he sold his car and used the proceeds to pay for his schooling at Embry-Riddle Aeronautical University in Florida—that got him his instrument rating.

In April 1949 he landed a job with All-American Airlines as a steward, a male stewardess as the job is mis-called these days. But stewards were common right after World War II, in fact they made up almost half of the flight attendant work force. TWA, Capital, Northwest, Eastern—they all had stewards in those days. And the smaller or feeder lines, such as Piedmont, All-American (now Allegheny), Mohawk, and North Central had stewards well into the fifties.

But Don worked as a steward only as a means to an end. He already had his commercial rating, and he was simply biding his time until a job as a copilot opened up with All-American. As things worked out, however, he got a call from Capital Airlines first.

"That was in 1953. I went over to Capital and worked as a copilot, flying DC-3's, for six months. You see, I had five years' experience with Allegheny [then All-American] as a steward, which meant I knew the airplane, the routes, and the lingo. So all of that helped."

But the job market then was somewhat similar to the job market today—there were more pilots than there were openings for pilots. Don actually had to leave the airlines for about half a year—for a stint with the IRS in Buffalo—before he was rehired by Allegheny and then finally recalled by Capital.

"So I was recalled by Capital, and I flew copilot for about six months, which gave me my year's probation as a copilot. But then, much to my regret, I was drafted into engineer's school [for training as a flight engineer, a nonpilot position; when the 'needs of the service' dictates, pilots can move up or, as in this case, back] and I went through six months of training in order to become an engineer on a Constellation. Now, no pilot likes to fly engineer, but it's better than not doing anything."

Conveniently, and because he was already a pilot, the downgrade to flight engineer meant an increase in pay. But there were other dividends, over the long run.

"In those days, you had to know as a flight engineer every nut and bolt and be able to discuss the whole engine, almost rip it apart and put it back together again in words. The engineer on a Connie really did a great deal of work in handling all the systems. All the pilot had was four throttles. We had the throttles, the [fuel] mixtures, the prop controls, the superchargers, the carburetor heat, and the pressurization. We had all of that back there on the engineer's panel.

"After three years as an engineer on a Connie, I was up for captain checkout, the check ride you are given by the airline to see if you are qualified to be a captain, which was what I was after. Unfortunately I was low on time, so I had to fly almost a year as a copilot again, to develop my time for captaincy. Finally I got my time in, checked out in a DC-3, got my Air Transport Rating, and started flying captain. That was in 1959.

"Then a few years later some of us got knocked back to copilot again, and I spent three more years flying copilot on Viscounts and DC-6's and Connies. 'Course you had to qualify [pass the company training course] on each one of those airplanes, each time you made a change. In 1965 I regained my rating as a captain and went through training school on both the DC-6 and the DC-7 and the Viscount.

"I flew those until the Boeing 737 came along. In fact, I was in one of the first classes of the Boeing 737 school, and I flew that airplane until 1972, when I checked out as a 727 captain, and I've been on those ever since. You see, on jets my airline will only allow you to be qualified as a line pilot, on one type of airplane at a time. You can't be qualified on two different airplanes at the same time in commercial service. If the company assigns you to fly a different jet, you have to go back to school and get qualified on it, even if you have flown it before.

"For example, I flew a 37 for five years, but I can't go back and fly a 37 trip now. I just have to fly my 727 trip now."

The airplanes—a Boeing 727 and 737—are very similar, but the company feels it is safer to have one captain familiar with one type of airplane, and one type only. So does the FAA.

"Things are the same way on the 37 as they are on the 27. You have the same engine, practically the same instruments, the flap speeds are practically identical, and a lot of systems are practically identical. It's a very easy changeover from a 37 to a 27. You've got one more engine, and they handle practically the same, but you still have to go to school to make the switch. They have the school down to about four weeks of ground school and simulator training.

FEAR OF FLYING

Then you go out and get some time in the aircraft, and then you have your rating ride. And that's it."

"That's it," makes it sound almost simple, but as Captain Don knows so well, there is nothing simple about holding down a job as a captain on a major airline. One fact absolutely should be known by every nervous air traveler, and that is that airline captains and copilots are continually being tested. And these are not gut tests that an experienced pilot can pass by rote; they are, to put it bluntly, ball-busters.

The FAA requires that every pilot be given a check ride every six months (and there is nothing simple about it, either).

The airlines insist on a pilot training flight with a flight instructor within six months of each check ride, regardless of the outcome. The FAA examiner can, and almost always does, ride along. With a company check pilot and an FAA man standing behind the flight crew, watching for the slightest error, the testing is an emotional experience, one in which a pilot is called upon to make maneuvers and procedures some of which he will rarely encounter when carrying the public. But the point of it all is that he must be prepared to deal with any emergency, from a fire anywhere in the airplane to a failed engine. And his response, based on both memory and his company flight manual, must be within stiff time limits. It is not uncommon for a PCR, a pilot check ride, to last over three hours.

If one of Don's fellow pilots flunks a check ride, he is given additional training and then usually allowed to take it twice more. If he still fails, he could be out of a job.

Modern technology has brought about one significant change in the tests in the last few years, which is that many of the pilot check rides are done without the crew's ever leaving the ground.

This is made possible by the use of the simulator—a marvel of engineering genius and mechanical skill, in effect an earthbound machine that flies. I know that sounds contradictory, but that is just what it is. The simulator is an exact duplicate of the cockpit of a particular airplane, and all the gauges, levers, and myriad other devices on the panel and above and behind the pilots are functional; they do just what they would do in a real airplane of the same design.

Immediately behind the crew area is an elaborate computer panel that the person giving the test uses to create simulated emergency conditions. If he wants to knock out one, or two, engines on takeoff or landing, he simply pushes the right buttons and the pilot is faced with exactly the same conditions he would encounter if he were in the air.

The more recent simulators are extremely sophisticated and can simulate wind conditions, rain, and almost any imaginable weather condition (the pilots who fly for Don's airline are now being given training and tests in flying through wind shear conditions—a direct result of the Eastern crash—in a simulator). A television camera attached to the machine focuses on a model city, laid out on a long table in front of the simulator, and as one pilot said, "You can fly right down between the buildings if you want to."

It may be difficult for the layman to believe, but within seconds of beginning his check ride, or training, in a simulator, even the most experienced pilot forgets that he is not actually in the air flying a huge aircraft. The simulators, jouncing this way and that on their elaborate hydraulic suspension system, are that close to perfection.

Don had just completed his six-month pilot check ride (in a simulator) a few weeks before we interviewed him.

"A typical pilot check ride for our six-month test? Well, first we take off under instrument conditions—we used to do this in the airplane under the hood [a device worn by a pilot so he cannot cheat and look out the window] but now we do it in the simulators—and they've given us the minimum conditions for taking off at that particular airport on that particular runway. For example if it were like Denver, it might be 2,400 RVR (Runway Visual Range) or half a mile visibility. They have us go down the runway and rotate. Just as you rotate, they will probably give you a fire in one of the engines, usually an outboard engine because that causes a directional control problem which you've got to correct immediately, there's no waiting. They give you perimeters for staying over the runway and maintaining your heading, and you've got to stick within them.

"Now you're leaving the ground, and you've got to make your climb out to 1,000 feet at V2 speed—that's the lowest speed you can make it at—and climb to that altitude with one engine down. You level off, pick up your speed, and make your normal flap retractions till you get your flaps all the way up and you're in a clean configuration. And you continue your climb on out to whatever altitude you've been cleared to, or to a safe operating altitude, again depending on the airport.

"At that point they usually program the weather to get worse so you can't go back in; the field seems to be closed now. So you have to start for another airport. But by this time we must have done all the proper procedures for an "engine-out," such as throttle closed, start lever cutoff, go to essential power, and so on. The man who is not flying does this, and he has to do it immediately and by memory, and then later at 1,000 feet the flight engineer reads the checklist to see that the copilot has done all this. And if he hasn't done what he should have done, from memory, he could very easily flunk the check ride.

"So you keep flying, and remember, you've got the company check pilot and the FAA man standing right behind you, and with that computer, they can give you any emergency they can dream up, and you have to deal with it properly and swiftly. But you have no idea what they're going to do next."

The emergency on takeoff is usually followed by some drills to make sure the pilot is able to do several steep turns, a holding pattern, and just when the crew almost begins to relax:

"Let's say they've got you up at a higher altitude now, maybe 33,000 or 37,000 feet, and then suddenly the standards man says, 'Ouch, there go my ears,' or something like that. There's no way to simulate sudden depressurization in a simulator because it isn't pressurized, so that's why he says that, or maybe he'll say, 'There goes a window.'

"You know you got to get down and get down in a hurry to 10,000 feet, where people can breathe without oxygen masks, so you begin a fast rate of descent. This is the emergency descent, or 'high dive' test.

"In order to get down fast, we pull the power off and pull what we call the speed brakes out. This disrupts the flow of air over the wings, and we really come down. We come down at 388 indicated knots. When you do this rapid descent, because of depressurization, the oxygen masks come out back in the cabin and, up front, we have five seconds to get ours on. We have to have that mask on and operating within five seconds of losing pressurization. You just reach over your head and just pull that mask right down

[76] CLEARED FOR THE APPROACH

over your head. There's a strict time limit there, so you have to hurry. Once in a while you knock your glasses off.

"What you're doing is pulling the power off, pushing the airplane forward, and reaching for the masks at the same time. You can't be smooth about it, you just get in that dive and start for the ground."

After the fast dive test, there are usually checks on how well the captain responds to fires, which could be of several types, and then the full crew is checked out in regard to electrical failures.

By this time the crew has been "in the air" for at least an hour. It has been, as the captain attests, "Intense all the time."

Next come the emergency procedure checks in regard to landing problems, and when these have been completed, the captain can rest, at least to the extent that now the copilot is tested on how well he flies the plane, and the captain assumes the copilot's role. In addition to all the above, the captain is also rated on how well he controls his crew during the many simulated emergencies.

Some idea of why the check ride takes more than three hours can be gained by looking at the items that must be covered, as listed in the airline's flight manual (and required by the FAA).

Under the flight proficiency section, there are more than a dozen maneuvers the captain must perform satisfactorily. Included are rejected takeoff; normal takeoff; takeoff with engine failure; terminal area departure or arrival; steep turns and stalls; nonprecision approaches; rejected landing and missed approach; holding; two-engine landing; normal landing; single-engine approach and landing; irregular and emergency procedures; and a no-flap approach. This is followed by a nine-item list for the copilot's proficiency check.

These in-flight (or simulated in-flight) proficiency checks are by no means the end of the testing. Once a year the FAA will monitor a pilot's performance by "flying the line" with him, which means that an FAA inspector will show up, usually unannounced, to ride along on one of the pilot's normal runs. But that is still in-the-air testing; there's a great deal that goes on down on the ground.

Once every year each pilot must take a written exam based on his company's operations manual, which goes into what pilots call legalities, meaning legal limits that govern such things as the altitude you must maintain when making a specific approach at a specific airport. The test is a multiple-choice, open-book test; it forces a pilot to relearn or remember a great deal of technical material.

The interview with Captain Don took place in the basement office of his spacious, comfortable home in the suburbs of a large East Coast city. The walls are filled with pictures of all the different planes he has flown, and the books on the shelves are almost all about aviation. (A copy of Ernest K. Gann's *Flying Circus* is a gift from his 9-year-old daughter, the youngest of his three children.)

In the bookcase nearest his desk are kept all of his company manuals. In response to a question concerning the annual written exam, he pulled out a thick binder and leafed through it, reading the sample questions and the multiple-choice answers. Then he stopped and frowned, "Here's a question that tests to see if you know what 'cleared for the approach' means. Boy, that makes you think about TWA 514, doesn't it!"

The question had to do with approaching an airport other than Dulles, but the correct answer—"C. maintain last assigned altitude until established on final approach course."—was, as Don put it, "right on point."

Pilots do not worry too much in anticipation of this exam because it is an open-book type, but they *do* exhibit some anxiety when it comes time for the Pilot Check Ride, because that is preceded by a tough oral examination where no help is given. It all has to be known or memorized.

(Don has made a tape of some of the more difficult points of information which he plays before going to sleep and even in the car as he drives to the airport. A random sampling of the tape turns up a wide variety of information so detailed and technical that one would think only a company mechanic would be required to learn it.)

The point of all this testing and training is to uncover areas in which a pilot might be weak, either in performance or knowledge. A slipup means more training or back to the books.

Even if a pilot's skills are superb and his knowledge staggering in respect to the amount of minutiae mastered, he can still be grounded because of his health.

The FAA requires pilots to pass two physicals each year and the company insists on giving its own, making a total of three. Most of the major airlines put their pilots through a grueling physical and disqualify men for any number of reasons.

* * *

Captain Don has come a long way since that first airplane ride at the Stockton Town Picnic. At forty-eight, he has a lot of seniority; he is among the top quarter of his company's thousands of pilots, and he gets the trips he bids.

Like most pilots, he's a bit sensitive to the complaints of people who think he has "too good a deal," and like almost all the airline pilots I've ever met—certainly the captains—he cannot bear to sit at home on his days off doing nothing. He sold real estate for a while, but now he limits his participation to the investment side. Yet when the need arises, he can be found on a homesite, hammering nails.

It should come as no surprise to anyone who knows pilots that on days when he doesn't fly and there are no nails to be pounded, Don is usually sitting around the hangar at a small local airport near his home. You can take the boy out of Stockton, but you can't take Stockton out of the boy!

At the end of the long interview, Captain Don summed up the extensive medical and physical testing and training, the many exams, and the seemingly-interminable checklists by saying, "I think mine is the only job in the United States where you have to prove yourself every six months."

That's just possibly why America has the best air safety record of any country in the world.

I wonder what the legal profession would be like if lawyers were similarly tested!

* * *

Airline crew members represent the visible part of the human factor that guarantees this nation its enviable safety record in the air. The *unseen* element is that of the air traffic controller. In this day and age, a pilot may be supremely skilled, but he must nonetheless depend on the comparable skill of his landlocked partner, the controller. Together—and only together—they make sure that you get where you are going without anything more serious than a minor delay.

Most people fly for years, both pilots and passengers, without having a good look at what an air traffic controller does, or understanding why he is absolutely essential to flight operations and particularly flight safety. In early 1968 when I met my first bunch of controllers, I doubt that one percent of the population would have recognized an air traffic controller if they had stumbled over one. Although the air traveler might occasionally glimpse the silhouettes of a few figures in the lofty glass towers that command major airports, the bulk of the work force was found in closed radar rooms and seldom seen.

Perhaps the function of a controller can be most clearly described as that of a *separator*. It is perfectly possible for a single qualified pilot to depart the East Coast, fly to the West Coast, land, and taxi to the ramp without ever talking to anyone. In its ability to find its way about, his airplane is an independent unit, taking its information from remote transmitting signals from the ground and moving from one to the next without outside help. To this extent, air traffic controllers are not primarily essential to that one pilot's safety of flight.

But should a *second* airplane depart the East Coast and head west, flying in clouds from time to time along the route, the risk begins. Although the odds may seem infinitessimal, the possibility arises that the two will meet in some slice of sky where the visibility is zero, and thus terminate both flights in disaster. When at any given hour of a busy day there may be *thousands* of airplanes in the air over the United States, many of which are plying identical routes through the weather almost simultaneously, the risk of a midair collision reaches the stage of probability. Without the "separators" working the system, the congested metropolitan areas would become a daily shower of twisted aluminum.

All flying, therefore, is divided into two different systems. The simplest, which accommodates most sport and pleasure fliers, is flight according to Visual Flight Rules, or VFR. Under these rules a pilot may take off from an uncontrolled airport, bore holes all over the sky, and land at some other airport without anyone ever knowing that he has been airborne. He may not fly near or through clouds, in conditions of poor visibility, or near busy controlled airports, nor may he fly below certain altitudes (that is, no buzzing) except to take off and land. To keep himself separated from other airplanes, he is commanded simply to "See and Be Seen."

For the serious traveler the limitations of VFR flight are usually intolerable. The uncertainty of arriving at a destination on schedule gave rise in the early days of aviation to the slogan "If you have time to spare, go by air," because all flying was at the mercy of the weather, that most unpredictable of all phenomena in human life. For that reason, all airline and most business flying is conducted under Instrument Flight Rules, or IFR, where it is assumed that the pilot will be in clouds from the moment he lifts off until just before touchdown. In these circumstances, visual separation is impossible, and pilots become completely dependent on someone who can "see" what they cannot: that someone is the air traffic controller.

As an example of what this dependence amounts to, let me profile a very typical flight from New York to Chicago in a Boeing 727. The captain has filed his flight plan with Air Traffic Control (ATC) well before the flight. (In fact, in all likelihood the airline computer has filed it directly with the ATC computer). As the flight is in its final boarding stages, he calls Clearance Delivery, announces his flight identity, and requests his routing. A controller then reads the clearance that the computer has furnished up, which may be exactly as filed or along a different route, depending on traffic conditions. Essentially, the pilot is told what departure route to follow, what altitudes to climb to initially and at what points, what his final cruising altitude will be, and what airways to follow en route to his destination. He will then be allowed to "push back" from the gate to start his engines.

As soon as they are fired up, he will call Ground Control for permission to taxi—and without that permission he is not allowed to move. Although it might seem reasonable to assume that pilots could find their way to and from the runway without chaos intervening, this would be a rather unsafe procedure at any busy airport.

In the first place, cockpit visibility in many planes (including the big jets) is somewhat limited, especially to the rear. Further, when there is fog about, a great deal of taxiing is done "in the blind" and requires a controller in the tower cab who is watching a "bright display" (a small radar set) to watch the ground traffic. Finally, without someone to direct traffic at taxiway intersections (there are no traffic lights) to decide who passes ahead of whom, huge airplanes would have to do a good deal of abrupt braking, which is hard on both the aircraft and the people in it.

When the plane has reached the active runway, the pilot switches to the tower controller (called local control by ATC) and waits his turn. Normally he will be first cleared into "position and hold" while landing aircraft roll out and leave the runway, and then "cleared for takeoff." A few moments after he has lifted off, the pilot will be told to "contact departure control," at which time he will be tracked on radar, then and for the duration of the flight.

Departure Control in the New York area is something special, for it refers to what is called the Common IFR Room.

Up until about nine years ago, New York's four major airports—Kennedy, LaGuardia, Newark, and Teterboro—were each governed by their own approach and departure radar rooms; each would *try* to coordinate what the others were doing by telephone. One day in late 1967, while climbing out of LaGuardia toward Morristown, New Jersey, I very nearly became a victim of this divided system. I was talking with LaGuardia Departure, and an Eastern DC-9 inbound to Newark was talking to Newark Approach. I popped out of a cloud deck at five thousand feet just as the DC-9 was descending into it, and I found myself counting the rivets on its belly as it passed less than fifty feet above me.

Controllers in the Common IFR Room now handle all approaches and departures for all four airports in a much more highly coordinated and safer fashion than was possible before. As our 727 climbs out of LaGuardia bound for Chicago, following the moment-to-moment instructions of the departure controller who can see it and all other aircraft in the vicinity, there is little risk that we are going to be bumped by a DC-10 inbound to Kennedy.

There is, however, one small risk in this phase of flight, which will recur during the letdown in Chicago *if* the weather is clear or only partly cloudy, and it is a risk that gnaws at pilots a bit. For even though we may be under "positive control" by ATC, our separation is "positive" *only* from other aircraft who are also under "positive control."

In other words, if the weather permits VFR flight in the area, there may be planes around which cannot be identified or even seen on radar, [because they lack the necessary equipment for sending an electrical signal to the ground] and the 727 flight crew will be forced to rely on "See or Be Seen" to avoid hitting them. Unfortunately, a number of midair collisions have resulted from this intermix of controlled and uncontrolled aircraft. To remedy the situation, the FAA is requiring more and more that all aircraft flying near busy airports submit to positive control, even if they are VFR.

After we have climbed a few thousand feet, our captain will be told to "switch to New York Center" and will be given the appropriate radio frequency. He is now talking to a center controller seated at a radarscope at the FAA facility in Ronkonkoma, New York, out in the middle of Long Island.

The New York Center has en route traffic jurisdiction for large segments of Connecticut, New York, Pennsylvania, and New Jersey, and a network of radar and radio stations located at various points throughout these states are all wired directly to the radar room. As the Center Controller makes his first contact with our Boeing, he is handed a "flight strip" which gives him our clearance and the route along which he is to direct us. And as we pass through 18,000 feet in our climb, both the controller and the flight crew relax a bit, for above this altitude no VFR aircraft are allowed, and the security of "positive control" becomes nearly absolute.

The entire jurisdictional area is broken into "sectors," each monitored at all times by at least two radarscopes: the low altitude controller and the high altitude (above 23,000 feet) controller. Because skyrocketing fuel prices are driving airlines up the wall, we will level off at 37,000 feet for cruise, an altitude at which three thirsty jet engines become somewhat reasonable. At that altitude (unless the captain requests a change in search of smoother air) we will pass through all of the sectors controlled by New York Center, and then be handed off in sequence to Cleveland Center, Indianapolis Center, and Chicago Center. About 100 or more miles out of Chicago we will begin our descent.

As we approach Chicago, at some point in our letdown between 10,000 and 4,000 feet, the Center will hand us off to Approach Control, which will issue compass headings (called vectors) calculated to line us up in sequence with other aircraft for the final approach to the runway. About five miles from the runway threshold, and just as we intercept the twin radio beams of the Instrument Landing System, we will be handed off to Tower Control for final clearance to land. Once off the runway, Ground Control will bring us to the gate.

I have assumed, of course, that the flight has been without a "glitch," as many of them are. On the other hand, many are not. Both the New York and Chicago Terminal areas are among the most sensitive in the world, and it takes very little to overload their normal operating capacity. We could have sat for an hour on the ground in New York because the Common IFR Room and New York Center were so glutted with airplanes that the radar just could not take any more. Or weather in Chicago might have reduced the operating capacity of O'Hare to a point far below the scheduled arrivals and departures. This could very quickly generate long lines of airplanes waiting to take off and stacks of airplanes grinding around in circles called holding patterns waiting to land.

That, at least on paper, is the world of the air traffic controller. It is not, however, a calm world, as I learned to my great dismay about nine years ago.

For the first fifteen or so years of my flying life, my firsthand knowledge of controllers was as slim as any other pilot's or person's. In fact during my military service, the only time I met an actual controller was when we were both assigned to investigate an accident! Although I talked to these faceless individuals every day over the radio, I knew none of them personally. And that is pretty much the way things remained until the late 1960s, when the air traffic controllers found themselves on the verge of a major labor dispute with their employer, the FAA, and went looking for a lawyer to represent them.

The call came from a gentleman named Mike Rock in late November of 1967. When I got the message, I had just returned to Boston from the west coast, and the note telling me to call "Mr. Rock of the FAA," startled me. To a pilot, a summons to call the FAA usually means a violation, a notice that he has broken some government flying regulation. As it turned out, however, it wasn't the pilot of 808 LJ (my Lear Jet) that was in trouble; it was the controllers themselves, at least according to Mike Rock, who turned out to be a very decent and worried young man.

[84] CLEARED FOR THE APPROACH

"Mr. Bailey," he said, "I called on behalf of myself and a few other controllers. You don't know us, and we don't know you, except for the conversations we have over the radio. We know you fly a Lear Jet, and we know the side number, but that's about it. I'm calling because we are in big trouble. Air traffic control in general has some serious problems, and as a result some of the guys are having individual problems. We don't have much money, but if we pool what we have I think we can come up with enough for a fee. We'd like to have you represent us."

Within a week of that phone call I met in New York City with Rock, Stanley Gordon, and a number of other controllers—about a dozen in all. At first, everyone seemed to be talking simultaneously, but after a while the discussion became clear, if not exactly organized. Their complaints shook down to these basic facts:

ONE. In the few years prior to 1967, air traffic had increased annually by 20 percent;

TWO. During the same period of time there had been no increase in staff, equipment, or other assistance; and,

THREE. Controllers were rapidly falling behind in their ability to perform the critical function for which they had been hired and trained—the responsibility for protecting airplanes and human lives.

They explained further that there had been two previous attempts at organizing (one a union and the other a professional association) but that neither had brought about the promised relief. They wanted to form a new organization, one that could articulate their problems in the right places and get some results.

I talked with them at some length and found that in the main they were former military traffic controllers, most of whom had a high school education, although there were a few who had attended or graduated from college and who had learned their trade while doing their stint in the armed forces. Upon discharge they had then been hired by the FAA and, after some training at the academy in Oklahoma City, had been given assignments of gradually increasing responsibility.

Pinnacle assignments were generally considered to be work in either the Chicago O'Hare area (the busiest airport in the world) the New York area (the most congested area in the world) or the Los Angeles airport area (which can compete with both when its problems are severe). The difficulties of function diminished somewhat, but not terribly significantly, as one moved to areas of less dense airplane population.

The average controller went to work for the FAA when he was in his middle twenties and, among the most distressing problems that were described to me, was apt to burn himself out by the time he was in his middle thirties. That simply means that he suffered something short of an outright nervous breakdown, but became unable to function as a front-line journeyman controller, because at somewhere between age thirty-five and forty, the intense pressures of the job became too much to handle on a daily basis. The New York Center, a large radar room located in Ronkonkoma, Long Island, near the McArthur Airport, was said to be hell on wheels.

In 1968 when I had my first opportunity to study the situation, controllers were using a rather primitive system to keep track of the identities of the various radar targets for which they were responsible. By using small strips of plastic called "shrimp boats" which were moved by hand across the scope as the target moved, the controller would try to keep in mind the plane's flight number, its altitude and whether it was climbing or descending, its routing, and its separation from all the other targets which were moving in different directions. To one watching over the shoulder of a busy radar controller and listening on a parallel headset, this was indeed a mind-boggling exercise. It required intense concentration and an excellent memory. The latter goes a long way to explain why air traffic controllers in general are a rather bright bunch of men and women—and they are —since one of the chief characteristics of high intelligence is a speedy and accurate memory.

From this view of the system, which I knew by rote from the cockpit but was now getting to appreciate from the inside, I had no difficulty in being convinced that the controllers' complaints were real, and the chronic overload had reached serious proportions. It was easy to visualize a shrimp boat marked "808LJ" moving directly at another shrimp boat marked "American 382," while an exhausted radar controller trying to supervise twenty-two other targets had momentarily overlooked the fact that both airplanes were converging on one another at exactly the same altitude. It was plain that these men were in need, very badly in need, of help from someone.

One of the controllers with whom I spoke at that initial meeting in New York described one incident where he had had thirty-five different airplanes on his radarscope (each one of which he had to memorize in order to keep track of them) and although he had managed to clear the scope without driving any two airplanes into one another, he had had nightmares for days thereafter, imagining what would happen if he mis-remembered even a single target long enough to put it at the wrong altitude at the wrong time.

I was told that controllers sometimes had to live on soup during the time they were at the scope, because they couldn't chew anything that would interrupt their ability to talk over the radio. They were unable to leave the scope during peak traffic hours even to urinate—they had to use number ten cans for that purpose.

One of the problems that was besieging these men was that a fatigued controller who wanted a break had to get someone to stand over his shoulder for at least twenty to thirty minutes so that the second man could absorb what is called the picture, that is, to have pointed out to him and then to memorize the identity and the flight strip of every single target on the radarscope. In other words, if someone were to drop dead at his position, there was no way that even an experienced controller who worked that position on a different shift every day of his life could step in and take over without a great deal of danger in the process.

For some reason, and probably because I am as much a pilot as a lawyer, the complaints of these men not only sounded very, very real, but also suggested to me that this was a cause that deserved an awful lot of attention in a hurry, from someone who knew enough about the system to intelligently discuss these problems in administrative and political circles.

It also sounded like a case which could be a substantial burden. The controllers naturally wanted to know what it was going to cost for me to represent them. I didn't have the heart to name a figure, because I learned that the average controller, at that time, was making just over eleven thousand dollars a year and suffering mightily, if he were in a high-density area, to even take home that paycheck.

The divorce rate was extraordinarily high, as is not uncommon in any high-pressure business.

And I rather fancied that the wives who were the recipients of the tattered and dog-eared controller, who left his shift and stumbled home, were having their own little nightmares to put up with.

I told the men that I was willing to help them, that it would be a labor of love, and that I would represent them for a period of six months (the maximum amount of time I could afford to devote to their cause) for the grand sum of one hundred dollars, plus whatever actual expenses might be involved.

During the year immediately prior to the request of Mike Rock and his associates, I had defended Dr. Sam Sheppard, Dr. Carl Coppolino twice, the Boston Strangler, Charles Schmid in Tucson, Arizona, Dr. Mitchell in Minneapolis, and John Kelly and Pat DiFario in the Great Plymouth Mail Robbery. It was, in all probability, the busiest year of my life, and had the controllers described their problem at any time before its conclusion, I might have had to lend them a less willing ear. As it happened, being on the back side of the crest of this wave of heavy trials, it looked like I would be able to give them the six months that they required and during that time devote a substantial amount of my own hours to their cause.

I should point out at this juncture that I am not and never have been or will be a union lawyer. I have worked with organizations a good deal, within the legal profession and beyond, but have always been a disciple of the independent view and not any real friend of collective bargaining or group action. Nonetheless, I felt that the proposition that these men were putting to me was completely legitimate and warranted some help. They did not wish to become a union, but were interested solely in a professional association that could effectively represent the voice of air traffic control and require, through sheer persuasion, some of the catching-up which was very obviously needed.

To be completely honest, I rather viewed the case as a simple one. After a few visits to the facilities where the controllers were working, and after studying a few numbers showing the increase in air traffic to be exactly as it was described to me, and after seeing that the FAA had lagged rather sadly in its duty to keep up with the need to provide more people and more equipment to handle the new airplanes that were coming into the system every year, I rather thought that a brief period of organization and some carefully described objectives would find universal support within the industry and with the FAA in particular, even though there might be a little egg on the face of that organization. I fully believed that within six months the controllers would be united in a common professional objective and that I could depart the scene at least having put the machinery in motion for ultimately solving the problem.

Of the number of mistakes that I have made in life—and there have been some whoppers—this has to come close to the top of the pile.

What I found, in essence, was the exact opposite of what I expected. And I suppose, in retrospect, that it sheds some significant light on one of the problems of the aviation industry as a whole, that is, a degree of conservatism that has no place in a business so fast-moving and as needful of the ability to anticipate a problem before it comes rushing over you as does the business of flying airplanes.

I began with a meeting with Mr. David Thomas, a career professional with the FAA, who was then the deputy administrator under General McKee, a retired air force officer, who was the number one man in the FAA. Mr. Thomas, however I might have agreed or disagreed with him personally, was at least a very honest man and freely admitted that the air traffic control corps was far behind its appropriate staffing and training, as related to the sharp increases in air traffic which were being realized in the booming years of the late sixties. He nonetheless failed to put a precise finger on the source of the problem; he thought the responsibility lay at the feet of Congress, which had failed to provide adequate funds. I rather suspected, and told Dave Thomas, that if congressmen thought that they personally were in one ounce of jeopardy as a result of their penuriousness with the FAA budget (assuming that was the true source of the problem) that they might well become very free spenders in an awful hurry, since most of them had to fly in order to exist.

I expected the support of the Air Transport Association, whose airplanes had to move on schedule in order to make money. But unfortunately the ATA (probably after some late-night checking with its friends in the FAA) refused to cooperate. Their attitude was akin to saying, "We know this industry. Why do we need you, an outside lawyer, to tell us what its problems are?" I'm sure the airline giants viewed the concept of a well-organized controllers' group as a spectre that might affect the bottom line of their profit and loss sheets.

The Airline Pilots Association was much more helpful and sympathetic, as befits those who are always in the foremost section of any plane that might find itself in danger of crashing. Some of its members stated that they would recommend strong support of any reasonable steps the controllers might take to promote greater air safety. General aviation also gave rather substantial support, however it was hardly with what one might call a single voice.

My own direct efforts on behalf of the controllers began with a meeting held near JFK Airport in January 1968. The meeting was billed as an experimental session wherein various proposals would be put to the assembled group for its reactions. I was surprised, not just at the size of the group, but at the fact that people had come from all over the country—including Alaska!

One thing was certain: these men were serious, and they wanted to start something more than just a social club. It was decided that the group would call itself the Professional Air Traffic Controllers Organization (PATCO). Although there was some skepticism because of the failure of the two previous efforts, the overall response was quite positive. A great many of those present signed application blanks that same night.

That meeting was held on January 11, 1968. During the next six months, I devoted a great deal of time and travel to addressing groups in other cities who had heard about the new movement and were very anxious to get some direct information as to where it was going.

By the middle of the year, when PATCO called its first annual convention in Chicago, the group was rather solidly organized. The convention was, in some senses, a backbreaking affair, because it was necessary to draw up what amounted to a constitution and bylaws for the organization, to set up the means of control through various rotating directorships, and to delineate very carefully what steps the organization would and would not support.

Most importantly, a safety committee was established to

[90] CLEARED FOR THE APPROACH

examine very carefully the practices which were being used at busy terminals to accommodate overloads, and the corners that were being cut in the process. To no one's surprise and in the confidentiality of the interviewing room of the safety committee, admissions began to roll forth that the intervals between airplanes was sometimes cut to as little as a mile in order to increase the number of flights that could be landed in a given time period, always racing just to stay even with a system that was bulging with a growth rate of 20 percent each year.

The FAA Academy in Oklahoma City (where the controllers are trained) had stopped recruiting during that period of time when it was apparent that air traffic was going to jump by leaps and bounds. As a result it was clear that the next two years were going to be tough ones. (Even though recruiting had started again, it takes a good two years before a new controller is capable of handling a sensitive position in a high density area.)

Each controller who reported infractions of this sort was asked by what authority the regulations, which clearly called for separation of three miles between airplanes approaching and departing the airport, were being violated.

The response was almost universal: everybody sort of looked the other way and just went ahead and did it.

Questions were asked about the number of close calls that had occurred as a result of shaving safety margins, which was a particular concern to me because during the prior year, while landing at San Francisco's International Airport and still in the clouds, I had suddenly felt a tremendous bump in the Lear Jet and had watched the angle of attack indicator go suddenly into the red which means that the aircraft is stalled. It rolled into a steep bank and I immediately came in with full power. I thank God to this day for the tremendous thrust which is available in the Lear because it is all that saved the airplane.

I dropped out of the clouds at 400 feet, well below the glide slope, and as forward speed increased got the Lear back to the point where it was flying once again and responding to the controls.

I looked ahead and there, less than a mile away, was a Super DC-8 not far from touchdown whose wake turbulence had obviously been responsible for nearly flipping my airplane in otherwise calm air.

The reports that intervals were being closed up beyond all reason were very credible as far as I was concerned. The upshot was that the controllers decided that they would issue a safety proclamation, which in essence was nothing more than a return to the regulations.

It was anticipated that unless the airlines responded and drew their schedules more realistically, there were going to be jams and consequent holding patterns at major terminals such as Chicago, New York, and Los Angeles. A notice was published of the controllers intent, in order to accommodate this kind of rearrangement, and was simply ignored by both the airlines and the FAA. I have to wonder in retrospect if the FAA really knew how bad its situation was and thought that the safety proclamation was a mere political device to beat the drum. In any event the enormity of their mistake very soon became apparent and there was little they could do about it.

Controllers began to enforce a three-mile separation on final approach of all airplane traffic and the over-scheduling very quickly overloaded the system and the delays became massive. The airlines' chieftans howled and demanded that the FAA discipline their controllers. The FAA could hardly discipline a controller for obeying the regulation, indeed it had no authority to invite him to ignore it or shave its requirements to meet the commercial demands of a burgeoning system.

A conference was quickly called at the round table conference room at Federal Aviation Headquarters at 800 Independence Avenue within the Department of Transportation. The problem was reviewed and FAA officials reluctantly agreed that they could not dispute the propriety of the controllers' action. Indeed they pretty much denied that there had ever been any shortcoming in the first place, although the circling airplanes over every busy airport furnished overwhelming evidence that the denials were frivolous.

I was heartened by the fact that General McKee, the outgoing FAA administrator whose resignation had been announced, pulled me aside and said "You know, why didn't you come around a year or two ago? I have been knocking the heads of those congressmen to try and get us the funds to make up for what we know is a terrible shortcoming ever since I have been the administrator, and I've been talking to closed ears. I think now that the correctness of my position has been demonstrated, that we will get some money and be able to bring this system back up to the strength and speed that it demands. I only hope that you can continue your efforts to help this group while keeping their professional standing on the high level which it seems to seek at the present time."

Gradually the air carriers knuckled under to the inevitable and began to schedule together somewhat more sensibly, having been granted authority by the Civil Aeronautics Board to confer at least on that objective. (Ordinarily they are prohibited from agreeing together on certain routes in order to provide the competition that keeps the service good and the prices within reason.) Under these emergency circumstances however, that kind of communication distance between individual air carrier companies simply could not be tolerated, and a waiver was issued.

Things went rather smoothly during the last half of 1968 and the early part of 1969. Needless to say, my original six-month objective faded like the wind, but the job continued to be, in my view, a worthwhile expenditure of time. Also, we had succeeded in appointing as a new executive director, a lawyer who had formally been with Lear Jet named Herman Meyer. At that time, I think he was the only other practicing lawyer who had a Lear Jet type rating besides myself. The organization continued to grow, its procedures began to shake out, and some evidence that a sincere effort to bring in new personnel and update the equipment began to appear.

The FAA finally allowed the organization to use the dues checkoff system, so that its income began to flow in a fair relationship to the expenditures that were being made for travel, publication of its magazine "The PATCO Journal," and other costs.

Late in 1968 when Mr. Nixon was elected President of the United States, Massachusetts Governor John A. Volpe was appointed head of the Department of Transportation. Although he had originally been promised the vice-presidency and was at the last minute squeezed out in favor of Spiro Agnew (an ironic loss to the United States, in retrospect) I was particularly pleased with the appointment of Secretary Volpe, because during the campaign I had from time to time flown him as a charter customer in my airplane as he stumped for Nixon.

Whenever he was in the plane, I deliberately left the loudspeakers on, so that he could become acquainted firsthand with the frenetic pace of the communications of air traffic control.

I had talked with other aviation organizations about the likelihood that he would be a good secretary and a supporter of aviation generally, and at the time of the announcement of his appointment he received some 4,000 congratulatory telegrams, the bulk of which came from journeymen controllers.

Late in 1968, just prior to his assuming office, I talked with him one day on the telephone and discussed the kind of man that he would want as FAA administrator. It was agreed that someone with expertise in aviation, probably not with a truly military background, but one who would be loyal to his own superiors, would have to be found.

Ironically, when John Shaffer, formerly an executive of Thompson Ramo Wooldridge (TRW Inc.) was installed in the post, Secretary Volpe had nothing to say about the matter. Indeed it was done when he was out of town, a move perhaps all too typical of the Nixon administration.

Soon after the appointment of Mr. Shaffer, I was brought around to his office for a courtesy introduction. A large and apparently confident man, his message in essence was, now that I am in office it won't be necessary for you to spend any further time with the problem; I will look out for the controllers and you need not be concerned. I was troubled by the sincerity of this approach. Of course, if he was telling the truth, he was in a far better position to really help air traffic controllers than I could ever be, and thus my efforts on their behalf would end on a very happy note. Or was he simply expressing alarm that he, as administrator, knew that he had very little control over the people who were trying to run the system and wanted it back so the annoyance from below would disappear? This was a most discomforting thought.

A few days later I got my answer. He had spoken about our meeting with an aviation leader from a different organization and had made it pretty plain that he wasn't going to tolerate any interference or much more noise from controllers—who ought to sit back and do exactly as they are told and no more.

There soon began to show up, in what had been a smoothly functioning system trying to repair its own needs, an attitude in management that suggested that someone at the top was saying "put the leg to PATCO." The controllers were patient at first but ultimately they began to react. There seemed to be management effort to encourage people not to support the organization, not to belong to it, and not to respect its wishes or suggestions. Meanwhile the recruiting program appeared to be bogging down.

One day in June 1969 I was sitting in my law office when my secretary said, "The head of the Department of Transportation is on the phone." I took the call and it was Secretary Volpe. "Lee," he said, "would you please tell me what the hell is going on?"

I said "Mr. Secretary, I haven't the slightest idea what you mean. Could you elaborate?"

"The system is all but shut down in New York and other major cities across the country are grinding to a standstill. A whole bunch of controllers have reported in sick, and I can't believe that it's a coincidence. If you were going to pull something like this you could have at least given me a warning!"

I said, "Mr. Secretary, please believe that I had no more notice of such a development than you did and, if it is in fact going on, it is not right no matter what the reasons for it, and I will come immediately to Washington and see what can be done."

I cranked up the Lear Jet just as fast as I could, since it was apparent that airline traffic had in fact been so disrupted that any chance of riding on a commercial air carrier was out, and sped on down to Washington's National Airport.

When I got there, the heads of PATCO rather reluctantly admitted that the sick-out had been the subject of some communications. Although most of the men claiming to be sick had good grounds in the sense that they were truly fatigued, the operation was not strictly kosher.

They had not informed me of what was about to happen for two reasons: one, they were pretty sure that I would disagree with it and attempt to block it; and two, they did not want me exposed to any punishment if it turned out that the government reacted with some sort of prosecution.

I could sense immediately that the move had been an enormous mistake. On the other hand, I could hardly abandon the controllers and come out publicly and condemn them.

I went to John Shaffer's office and was informed that he

had been ordered to confer and end the walkout just as rapidly as possible. I told Mr. Shaffer that I disagreed with what had been done, and that in exchange for *his* agreement not to penalize any of the controllers involved—in other words, a grant of immunity of sorts—I would appear publicly and call upon them to return to their jobs as swiftly as possible, except for those who were totally physically unable to do so.

He gave me his word, and I made the horrendous mistake of not requiring that he put it in writing, because it turned out to be a most volatile assurance.

I immediately called a press conference and asked the controllers to come back to their jobs and informed them that their problems would be the subject of negotiation at the highest levels of the administration. They responded, somewhat to the chagrin of the leaders who felt that a serious mistake had been made, and within a short time the administration began to prosecute large numbers of them, claiming that they had violated the regulations.

As a punishment for their actions, the dues checkoff system (whereby organization dues were withheld from the paycheck and forwarded directly to headquarters) was terminated. The purpose of this move was to throw the organization into financial disruption, hopefully killing it entirely.

The summer of 1969 was a long and hard summer, but ultimately, despite some falloff in membership, PATCO held together. Once again however the acrimony between the FAA management and the journeymen controllers who ran the system began to build internally. It seemed that every effort to frustrate the objectives of the organization and its ability to function was being made in the most nit-picking fashion.

PATCO members were not allowed to post notices of meetings in the FAA facility and were denied any recognition whatsoever. If ever a bureaucracy demanded that it be confronted with a problem, the FAA took each necessary step to guarantee that result, and by December 1969 the system, insofar as its essential personnel were concerned, was on the verge of coming completely apart.

By early 1970 the board of directors of the organization had determined that the only way for it to fight back was to take every controller who had a legitimate medical complaint and encourage him to enforce that complaint by reducing his services accordingly.

In fact, a large number did have disabilities which would have warranted a thirty-day layoff or better, as was ultimately shown by formal medical examinations.

At this point I felt personally that I was in a very precarious position, since any job action beyond that which could be legitimately grounded on medical reasons could be construed as a strike against the United States. And that was flatly illegal. On the other hand, to the extent that the board of directors was correct—and there was no doubt that they were correct to some degree—it would be necessary to give appropriate legal advice as to what they could and could not lawfully do.

I gave serious thought to withdrawing at that time simply because holding a license to enforce the law and nonetheless participating in a knowing infraction are to my mind entirely inconsistent—even though a great deal of good has been done in this country in crises by what is properly called civil disobedience. (That is, simply, peaceable demonstration in order to move some great arm of government that has proved too lethargic to do what it is supposed to do.)

I crisscrossed the United States in the Lear Jet, meeting with large groups of controllers and making plain to them that as a lawyer I had to advise them that they had no right to agree together as to who would be ill or who was ill, and that each man had to make a personal decision and he had to make that decision based on some genuine belief of illness. To do otherwise would be unlawful. I advised each controller to visit a physician, report his complaints, and ask for medical advice in the normal and routine fashion, rather than simply sitting around the house saying that he was too tired to work.

It was, in a sense, the most diffuse kind of leadership, but it took some of the impact out of what might otherwise have been a complete and total shutdown of air commerce in the United States.

All during this tour, I was constantly on the phone with the officials. I had very little to talk about with John Shaffer at this point, since I could not, in my own mind, take his word again. I did talk at length with Deputy Secretary of Transportation James Beggs who, although very hard-nosed when it came to controller problems, was clearly fearful that the administration was on the losing end of a situation it could not effectively fight and was in a panic to keep from getting egg all over its face.

I suspect that Secretary Volpe (who had found to his chagrin that he had very little control over the people who worked under him since even at the highest levels they were protected by civil service regulations from being discharged) was somewhat amused by the entire matter and strongly tempted to say "I told you so" to the people who had insisted on attempting to solve their problem through internal warfare rather than reasonable negotiation.

In any event, the night before what was believed to be a deadline, even though nobody called it that, I was with a large group of controllers in New York City when a call came in to have an emergency meeting with the Secretary of the Department of Transportation the following day, a Sunday.

We arose early in the morning, only to find that the Lear Jet was covered with about three inches of ice and in no condition to fly. We hauled it into a hangar at McArthur Airport and began the tedious business of washing down and removing the ice as rapidly as we could; some two and a half hours later a number of the key men from the organization and I were finally in the air.

We sat down and had a tense but I thought a productive meeting with Secretary Volpe, Mr. Beggs, Administrator Shaffer and others in the hierarchy. It was determined that the principal source of rankle (and in fact this problem was no more than the straw that broke the camel's back) was the impending transfer of three PATCO leaders from the Baton Rouge, Louisiana, tower to remote areas around the country. This move was occasioned strictly because the management could not tolerate the voice of PATCO, and the organization was to be broken up forcibly by this method.

I warned the secretary that I did not have control over what seemed likely to happen, principally because of Mr. Shaffer's breach of promise in May 1969 which had made it apparent that I could not successfully and reliably deal with the FAA. Therefore, even if I had been able to go on all three networks at once and demand that the air traffic controllers abandon their plan, I would have probably been ignored. I could only help, I felt, by attempting to negotiate the differences and find some way to relieve the immense pressure that was building up.

It was suggested that the Federal Mediation Service, which is called upon continually to attempt to negotiate impending strikes where the commerce of the entire nation can be affected (railroads, auto workers, etc.), might be a proper middle ground.

The PATCO directors who were earnestly seeking a solution other than one of confrontation very quickly agreed, and a mediator was appointed.

Although it seemed as if the day had been successful, I warned the directors that one other result might develop, and that was that the administration leaders were simply buying time so that they could organize in an effort to defeat a sick-out, should negotiations break down and one eventually occur. That is, they would simply attempt to retrain all of the controllers in management who were no longer fast enough to work the board, as the radarscopes are called, and also to arrange for young recruits to step in on an emergency basis and perhaps do work for which they were not qualified.

The hearings began before the federal mediator Kenneth Moffet and it became almost immediately apparent that he felt the controllers had a pretty good case. After hearing a number of witnesses and presentations from both sides, he recommended to the Department of Transportation that the Baton Rouge controllers be left right where they were for a considerable period of time while further negotiations took place to attempt to get at the root of the problems. For it was apparent that the boys in Baton Rouge were merely the trigger on the gun.

Two months had gone by while all of this took place. The men at PATCO were quite pleased that their demands had been vindicated by an impartial source, and therefore were more than shocked when the FAA announced it would completely ignore the mediator's recommendation—and ordered the transfer of the men from Baton Rouge.

The bureaucrats well knew that in so doing they were asking for the battle that had been averted in February, and it became immediately apparent that they had prepared as best they could.

In what was almost a roar of wounded defiance, the PATCO directors all over the country condemned what they labeled a bad faith move by the administration, and within twenty-four hours fifty percent of all the controllers at work had reported in sick. This happened most heavily in major areas.

The government at this point had no desire to negotiate but only to fight, believing that they would win. Indeed one of Secretary Volpe's key men, with whom I had been friendly for a number of years, talked with me confidentially the night before the men left work and said, "It looks like an even battle, and we've got the upper hand. The secretary says 'Good luck, and let the best man win.'"

The Justice Department, through Bill Ruckleshaus, the head of its civil division, immediately brought suit asking for an injunction against PATCO in order to stop any strike. On behalf of the organization, I went down with its president, Jim Hays, and the chairman of the board of directors, Mike Rock, and agreed to a judgment to that affect, that is, that no controller had the right to strike and anyone who was absent from work merely for the purpose of striking should return. I made the appropriate announcement on television as to what the judgment had been.

Needless to say, as in most cases where a federal court attempts to control by order a job action by a bunch of highly unified and emotionally upset people, there was very little visible effect. In fact, within the next twenty-four hours the absenteeism began to increase. The centers at Boston, Cleveland, New York, Chicago, Kansas City, and on the West Coast were badly hurt and able to control far less than fifty percent of normal traffic.

For those planes that could get out of New York, it was necessary to come down through Washington and across the southern half of the country in order to get through the system to the West Coast. Atlanta, Washington, Jacksonville, Memphis, Dallas-Forth Worth, and Albuquerque still had a fair number of controllers at work, many of whom were PATCO members who could not honestly say that they were so fatigued that they were entitled to a few days respite. Air traffic in those quadrants limped along.

The FAA was desperate to have the public believe that the controllers' efforts had failed and that there was not sufficient unity in their ranks.

We soon discovered that as part of that plan, my predictions of sixty days earlier were becoming reality. Management personnel who had been journeymen controllers years ago, but were hardly current to perform the job, were stepping up to the radarscopes and moving traffic. Sitting beside them were trainees who had neither the experience in the system to be qualified to take front-line responsibility for the movement of airplanes, nor the standing in the organization to be protected if they had joined with the other controllers, nor any real justification to claim illness because of the short duration of their tenure and their youthful good health.

The fact that collisions did not occur left and right was due to the great diminution in air traffic—that is, there simply weren't enough airplanes to present the intensity of traffic which could lead an inexperienced or rusty controller to drive two of them together. Aside from that saving grace, it was one of the most reckless steps that a government agency has ever taken. Indeed, it was tantamount to using retired airline captains who haven't flown in some years and new trainees who haven't qualified as flight engineers to attempt to keep airplanes in the air, if all the regular line officers refused to fly.

In addition to everything else, this situation put intolerable burdens on those controllers who *were* qualified. An unqualified controller could easily make such a mess of the air traffic under his aegis that when he handed control over to someone more experienced, the qualified man wouldn't have enough time to sort things out before an accident occurred. I thought the qualified people were obliged to refuse to work under such circumstances, much as an airline captain would have to refuse to fly if his copilot were not qualified (as the regulations wisely require).

In a broadcast about forty-eight hours after the walkout began, I strongly suggested that in every facility where these two conditions, or any combination, were causing legitimate controllers to violate the regulations on an hourly basis, they should refuse to do so and leave the facility rather than risk a collision put in motion by unqualified personnel.

While this broadcast had virtually no effect on the numbers that were out—and these numbers had by this point hardened into firm lines of resistance—the Justice Department took umbrage and immediately filed a petition that Mike Rock, Jim Hays, and I be held in contempt.

A hearing was set for Monday morning. I assembled the names of a large number of controllers who were not scheduled to be on duty during the court hours that we anticipated and subpoenaed some ninety of them to appear and tell the judge the reasons justifying my statement and the probable consequences of the FAA's continued attempt to cover-up the extent of the crisis. (In retrospect, it seems that "cover-up" was a policy which will forever mark the administration of Richard Nixon.)

At about 10:30 on a Sunday night, a United States marshal friendly to our cause came to my door at the Mayflower Hotel and informed me that the Justice Department had just gone to the home of Federal Judge George Hart and told him that I was out of town, but that I had subpoenaed all working controllers from an *active* shift as a ploy to shut down the Washington Center and Washington Towers and asked that he quash all the subpoenas. Accepting their representations as officers of the court, he did.

At that time the acting attorney general of the United States was Richard Kleindienst, with whom I enjoyed a pretty good relationship despite his somewhat hard-nosed attitude about the issues involved. (It is a singular piece of irony that some years later that same judge would have standing before him former Attorney General Richard Kleindienst entering a nolo plea to a misdemeanor.) In any event (the memory is very much like it occurred last night) as a result of that tip, I was more than boiling mad, but not distempered, simply figuring that the government had pulled a bold maneuver by misrepresenting to a federal judge at his home that I wasn't in town, when in fact I felt very strongly that they knew I was in town. I say this because I had been informed that they had a tap on my line at the Mayflower Hotel and would have good evidence of my movements if I ever left the room, which I had not. I had asked my then-partner, Gerry Alch, to come down to represent the president and the chairman of the board.

Together we consulted one of the leading lawyers in Washington, who informed us quite correctly that he didn't know enough about the air traffic control system to intelligently try the case and couldn't learn enough in less than thirty days (and probably sixty) to do so. He offered the gratuitous information that Judge George Hart was one of the toughest judges in the District and would undoubtedly put me right behind bars.

Although I had no particular desire to go behind bars, I

felt that any judge who would take such strident action in a labor dispute might well, by that act, assure a degree of success which neither their leaders nor their rather reluctant counsel at that point had any right to contemplate.

After a series of inquiries I found the home number of Judge Hart and placed a call to him. I asked him, in terse but respectful tones, whether or not it was true that he had in my absence issued an order quashing subpoenas of all the witnesses I felt were important to the hearing the following day.

He said that he had done so on the representation that they were working controllers on shift at the time of the hearing and that I was unavailable to make comment.

I told him that I was very much available and that the government had simply not disclosed that fact to him—as perhaps they should have.

I should point out that in all probability Bill Ruckelshaus was not party to the deception if any was involved, since he has been and continues to be in my opinion a pretty straight guy.

In any event, the judge assured me that if any prejudice had resulted from the elimination of the ninety witnesses that I thought were important to the issue before the court, and it was strictly one of contempt, that he would see to it that no injustice was done. In other words, implicitly I thought, he would simply postpone the hearing—which is about the last thing the government wanted—and reinstate the subpoenas if he were convinced that he had in any way been deceived.

I began to feel somewhat more comfortable about the attitude of Judge Hart than I had when my lawyer-acquaintance had described him initially.

It then occurred to me that if the United States government through its Executive Branch had had the temerity and the gall to throw a tap on the phone of counsel, it probably had listened in, willingly or otherwise, to a conversation between counsel and a United States district judge. Playing the probabilities, as my colleague Louis Nizer once recommended lawyers do, I called in my good friend the marshal and sent him on a midnight errand.

I asked him to serve subpoenas, in shotgun fashion, upon the FBI, the FAA, the DOT, and any other agency that we could think of, requiring that they bring to court the following morning at 9 o'clock all transcripts, tapes, witnesses, agents, and other evidences of taps upon my phone.

In the first place, I was very interested in finding out the extent to which they were tapping me, and I had no doubt that they were. (In light of recent developments, I *shouldn't* have entertained any doubt.) In the second place, I thought it might be of interest to a United States district judge to learn the enormity of the government's invasive powers, in light of their very limited rights.

The following morning there was a substantial crowd assembled in the courtroom of Judge Hart.

I made an immediate motion that inasmuch as I viewed the matter as one of criminal contempt, that we were entitled to a jury trial and should not be put to litigation before a judge sitting alone.

Judge Hart had very little sympathy for this proposition, suggesting perhaps correctly that this was strictly an issue of civil contempt which could be purged any time I chose to retract my position, if in fact I were proved to have been wrong.

Without further arguing that issue, I asked that the government be called upon to furnish the wiretaps that were called for by the subpoenas which they had received, albeit in the middle of the night, but prior to the hearing. Carl Eardley, the chief of litigation present in the courtroom on behalf of the Justice Department and a number one aide behind Bill Ruckelshaus, chief of civil division, responded (and I must say Carl Eardley was an honest man from beginning to end) that he "could not say" whether wiretaps existed or not.

Sometimes a lawyer's life hangs in the balance of his ability to read a judge. I may have been right and I may have been wrong, but I rather thought I saw the hackles rise in the back of the neck of Judge George Hart, who was indeed as he had been described, a considerable man but not in my opinion at that juncture a hangman. Whether he articulated the matter or not, I'm not entirely sure, since we never had cause to have the record transcribed and no appeal was ever taken, but I am perfectly satisfied in my own mind that he was more than disgruntled at the possibility that a United States District judge might have been the subject of a federal agency wiretap. And I am more than satisfied that the record shows that at that point he said, "I suggest that counsel meet with me in my chambers."

I should like to disclose the conversation which then took place. I think that under the circumstances since I was treated as counsel in those chambers as I have been before and since I will not give chapter and verse as to what was said and by whom. But I should say that both Mr. Ruckelshaus and the federal judge in charge of the case took a very realistic view of the problem which confronted them and began to seek realistic solutions. Indeed, for a long half hour I had a rather intense pride in the law at work and a rather strong hope that a practical conclusion would be reached.

I cannot make the same statements for counsel who appeared for the FAA and the DOT, but I do not think that Judge Hart will consider it a breach of the confidentiality which must obtain in judicial chambers during negotiations if I mention the fact that he quite wisely made it plain, implicitly or explicitly, that Mr. Ruckelshaus return for further conferences without his clients, whose obstreperous demands were both unrealistic and unlikely to be productive.

In other words, to put it bluntly, they suggested that I be put in jail at once as a solution to the problem, something that Judge Hart could see would only aggravate it and probably by a considerable degree.

Ultimately there was a trial. The trial consisted of my conducting a case attempting to prove that what the controllers had done was the product of genuine sickness. Certain medical experts were produced to describe what they had personally seen in the Kansas City Center, that is, mature men who were so completely broken up by the whole affair and the probable loss, as they viewed it, of their entire position in society as air traffic controllers. Whereas they had been very structured proud, accomplished individuals who had been so highly trained for what they did that they were not easily adaptable to any other form of labor, they were now on the very brink of despondency and certainly in no condition to control airplanes.

The doctor was a Dr. Wayne Sands from Des Moines whom Verne Lawyer had gotten to handle the matter. The chief trial lawyer for the government was Carl Eardley, who did a yeoman job in the face of almost complete unawareness of the way the air traffic control system worked.

At one point I found it necessary to take the stand as a witness for the organization. Verne Lawyer, deputy general counsel to PATCO who had been present throughout the proceedings strictly as a labor of love, then took over the questioning, since I could hardly be both a witness and the attorney in the proceeding. Judge Hart, I must say, while keeping a rather stern visage throughout the trial, nonetheless showed a substantial interest in the text of what was being laid out before him and a rather liberal attitude for one who had been vouchsafed as an irreparable conservative, in allowing the rules in court to fluctuate as the needs of the moment might dictate.

I gave my appraisal of the situation and also my view of the statistics that supported the notion that without cheating the FAA could not possibly have maintained its alleged staffing. At one point Mr. Eardley, who tried valiantly and professionally to uphold his end of the Justice Department's burden by making informed inquiry of all witnesses on the stand, put it to me cold turkey.

I had made the statement that New York Center, based on what I knew about the absentee rate, was handling more than 50 percent of the airplanes (or less than fifty percent as FAA statistics might reveal), but with far less than fifty percent of the help that they required for such numbers. In other words, whatever was being done, more airplanes were moving through the system than the number of experts who were actively working the boards would permit.

Mr. Eardley circled me like a hawk (as good trial lawyers often do) preparing to pounce with a point which I am sure he felt was determinative of the issue that I was pressing. In essence he perceived that if TWA's pilots in 707's leave the area and TWA can make a deal with Pan Am to put their 707 pilots in, then the system will roll. He missed an important point.

He said, "Mr. Bailey, how can you be so sure that when the New York controllers you say are absent left their posts, *we* didn't bring in a bunch of controllers from Chicago's center which is equally busy and begin to staff the positions?"

It was a legitimate question from a lay point of view, but one which should have never been put to someone who knew the system. I said, "Mr. Eardley, the answer to that is twofold. First of all there aren't enough people left in Chicago who would buy such a deal, who would make such a move; and second, if they did, it would take them six months to legally be checked out in positions for which they were fast enough, but had not the familiarity necessary for being licensed to operate."

There are a few times in a lawyers' career when one sees counsel absolutely stopped in the middle of the courtroom, groping for another question after having been destroyed by a response. If I had my druthers, I would visit such misery upon someone that I either profoundly disliked or who had no business in the courtroom to begin with. Brother Eardley was simply in over his head, and although I'm not sure that Judge Hart participated in my quiescent glee to the degree that I did, he nonetheless got the point.

The upshot was that he found that because of the overwhelming numbers of people who were absent, that there was in effect a strike. I am sure that as the government read the first sentence of his judgment, they were jubilant, thinking that they had won the case against PATCO, for whatever value that might have. I'm sure the remaining sentences were less rewarding. In effect, he said yes, the air traffic controllers went on strike—but it was the result of extreme provocation by the FAA. This is tantamount to saying, "Okay, he committed a homicide, but with extreme provocation; he varied somewhere between temporary insanity and not guilty." It was a condemnation of the FAA which in all probability they didn't understand, but certainly the wiser and cooler heads in the Justice Department understood fully.

Meanwhile, another tack had been taken in what was to become one of the monumental interstate battles of the decade, and it was only the threshold of the decade, four months into 1970.

The Justice Department decided that inasmuch as its suits against the corporation were not having any visible effect on its objective (which was simply to get all the boys back to work) that it would sue all the controllers individually.

Once again it was a strident tactic, I rather suspect one dictated by the White House or some bureaucratic official of short depth perception (certainly not enough for the FAA to allow him to fly an airplane) to bring simultaneous suits in sixteen federal districts against those controllers individually who were not reporting to work. Of all the poor strategies (and it hardly deserves that title for it was an off-the-cuff tactic much as the fellow who shoots from the hip when he's too far away to find his target) it probably determined the outcome of the case at the very moment the suits were being filed.

I can still envision the federal secretaries across the country, furiously typing long petitions asking for restraining orders, and working with address cards in order to get each and every offending air traffic controller as a defendant in the suit.

There were two problems that the government did not see as these suits were filed. Number one, if you sue an individual he has the right to be in court during the proceedings, and one can hardly get an air traffic controller back on duty during the day if important litigation is pending against him which causes his absence from the FAA facility and his presence in court. Number two, United States attorneys across the country, however commendable they may be in other directions, knew little or nothing about air traffic control, whereas PATCO had accumulated many lawyer-supporters (largely through the goodwill which controllers generated to pilots who fly, many of whom were also good trial lawyers whom I knew and could call on).

In fact, I had at my disposal a rather substantial team of what I like to call nut-cutting trial lawyers, those who can jump into any litigation and based on habitual knowledge put forth a commendable effort even on short notice.

The results as to the government's case were simply disastrous. The government had expected the controllers, after having been named individually, to come forth shamefacedly with their tails between their legs begging for an opportunity to go back to the radarscopes and control whatever air traffic might wander through their areas. In every such case the opposite occurred.

Tough World War II and Korean pilots who at that point knew more about the law than the Justice Department found the gauntlet dropped in their laps. They responded by marching into the many federal courts where actions were commenced and putting on defenses and bringing motions which the government could neither understand nor comprehend and could certainly not deal with.

Everything came to a screeching halt, as if Mr. Freeze had sprayed the entire operation.

[108] CLEARED FOR THE APPROACH

It did not take the Justice Department long to determine that it had made a fatal mistake.

At one point, Judge Hart ordered me to tell all able-bodied controllers to go back to work. It was a perfectly proper order and I attempted to comply with it.

But vast numbers of controllers were not at home. Instead, many of them could be found in federal court, as the direct result of the government's ill-advised decision to name them as individual defendants.

I remember as if it were yesterday one extremely dramatic scene that took place in federal court in Brooklyn, New York.

With the more than able counsel of two fine trial lawyers, Marvin Segal and Jim Catterson, the controllers were embroiled in a gigantic suit that included as parties the United States Air Force, the Air Transport Association, and the FAA, among others.

The judge upon whom this holocaust was visited was Oren Judd, a jurist of marked ability before whom I had appeared several times previously. He was determined to run an orderly trial, one where everybody would be heard, despite the almost panicky complaints of the FAA and the Air Transport Association. My sole function in the proceedings was to step forward and tell Judge Judd that I had been ordered by Judge Hart in Washington, D.C., to inform all air traffic controllers within earshot that if they were not sick they were thereupon to return to work.

(As long as I was there, I took advantage of the opportunity and asked Judge Judd that I be severed, or separated, from the case—for reasons still unknown to me, the Air Transport Association had seen fit to sue me along with the controllers.)

What made the courtroom in Brooklyn an ideal spot in which to deliver the message was that some 275 controllers were present as a result of having been named defendants in the case.

At this point, however, the question of who was or was not sick had been mooted by the United States itself, because everyone—however sick or well—had a constitutional right to be present and to listen as his case was presented. This circumstance was not beyond the cognition of Bill Ruckelshaus, who continued to carry a cool head, and I am sure that both he and Judge Hart were fast coming to the conclusion that this was one of those cases where federal litigation is not necessarily the answer. The court is seldom a be-all and end-all for, or even an effective control over, job actions involving employees who are genuinely upset by the conditions confronting them.

(I indicated at the outset of this chapter that I was not, have not been, and do not intend to be a labor lawyer. Indeed, prior to my experience with PATCO, I had a distinct and unequivocal prejudice against that type of representation and the philosophy it embraced.

But I did learn, perhaps for the first time in my career, that there *are* situations where individual action will never be sufficient, and that men and women have to band together in order to raise an effective voice concerning a serious problem.

In a sense, this case was a golden opportunity for me to learn that lesson, for the PATCO case was a selfless issue in the controllers' minds, and therefore one without any malicious intent behind it, but simply a deep concern for the system they were charged with running. To see these dedicated men running out of steam before they reached their forties—the time that we doctors and lawyers generally feel is when we reach the expert level—substantially altered my views as to the rights of labor and the obligations of management.

I don't lay claim to an absolute conversion, but I do feel that the PATCO experience has helped me considerably in the management of my own affairs, especially in regard to my own employees.)

In any event, there was a general agreement by the Justice Department to abandon all the suits it was losing (or those that were being stalled on a daily basis) provided all the judges involved would agree. And all did except for one.

And that judge was Thomas Lambros, of Cleveland, Ohio, then, at age thirty-nine, the youngest federal judge on the bench. When Judge Lambros was practicing law, I had known him, and we had enjoyed a good relationship.

His attitude was, this matter has been placed before me, I'm called upon to resolve it, and I shall do so. As I listened to the report come over the telephone that there was a lone obstacle to at least depressurizing the raging controversy my heart sank. I thought, one federal judge, indeed, a fellow

I have known is submerging what we have at last agreed together as professional lawyers might be a sensible way out.

And not only was Judge Lambros threatening to upset the whole applecart, but Cleveland controllers were reporting that he gave them a rather severe tongue-lashing in open court, bringing several wives to the point of tears. What's more, he apparently included me in his manifest unhappiness. It looked as if I had better get to Cleveland right away.

That, however, presented a problem. All during the walk-out I had been flying my twin-engine Cessna, which was more practical on low-altitude trips than the Lear, but only during VFR conditions. (I couldn't in good conscience contend, on the one hand, that the system was being manned by rusty former controllers or raw recruits, and on the other hand file to fly within it, thereby admitting faith in those same controllers. Frankly, as a pilot I wanted no part of it, because the reports filtering back about who was doing the work were truly frightening; I felt far more secure on a see-and-be-seen basis than I would have under what was left of the air traffic control system!) Nonetheless, Verne Lawyer —a most accomplished pilot—and I climbed in the Cessna and somehow managed to avoid those clouds that would have required us to file a flight plan.

Things began to pick up shortly after our arrival. I told the news media that I hoped the judge would look carefully at our side of the issue, and then learned that he had already begun to do just that. He had gone to an FAA facility and he'd taken a pretty good look at what was going on.

Indeed, he had changed his point of view, or at least the point of view the controllers thought he held, and ordered that every air traffic controller claiming to be sick be examined by an independent panel composed of experts who were neither slavish to the FAA nor agents or friends in any way of the controllers, but simply impartial. (It was a wise judicial tactic which was to be repeated many years later in a case of perhaps more public interest—Patty Hearst.)

In effect, his solution brought about the termination of what could have been a disastrous strike, and by "strike" I mean an action by people who had become so emotionally upset by what was happening to their compatriots that they simply departed the job.

In fact, and although it was never taken seriously by the

one of them. Indeed, throughout the walkout you have been on the couch in effigy in FAA headquarters."

I thought that to be a rather small tactic for people who should be concerned about air safety, but a rather large tactic if they put the focus on the one person who probably could have led (if he were willing to turn in his license to practice law) a walkout large enough to cripple the United States and produce an effective demonstration.

I'm not sure to this day what was right or what was wrong. I did resign according to the request of the air traffic controllers. Not that they were ungrateful, they never had been, but simply because it was the thing to do at the time it was requested.

I came to PATCO as an aide to an idea. I left PATCO as an established organization. Albeit, it was the product of the efforts—the full-time efforts—of many people other than myself. For the first time in my entire life I had not been able to focus on a client as an individual. But I was afforded a peculiar insight into the group I had heretofore been in quasi-contact with because of my intimate involvement with the disembodied voices that said do this and do that, change this frequency, and then do that. These strangers across the country became real men. I can understand controllers as I think perhaps no other active pilot has ever been able to do, because I stuck with them, agonized with them, tipped a Scotch with them, and listened to their problems.

We became, I suppose, rather close friends on a wholesale basis. I watched men who had no training for the positions become executives, directors, thinkers, decision-makers. Although when we first met they had the lowest profile in aviation, and although there had been alternatives, there should have been some better way for the public to learn that the air traffic controllers existed. Nonetheless, with ten years' hindsight, I think the effort was both appropriate and worthwhile.

What was accomplished by all of this? A great deal, I think, although it certainly came painfully and slowly.

A great many new controllers have been hired and trained, reducing the overloads which used to be carried by too few. A large portion of the radar system has been upgraded and computerized, minimizing human error. The old "shrimp boats" have pretty much given way to little electronic tags which travel along the radarscope with the target, constantly reporting its speed, altitude, identity and other information which the controller used to have to carry in his memory. And although it was not one of the initial demands, the income of the average journeyman controller has just about doubled, bringing him somewhere near the fair wage to which an expert of that caliber is entitled. Finally PATCO, which was born as an idea just over eight years ago, today commands a higher membership percentage than any other organization of federal employees.

I did not make money representing the air traffic controllers of the United States and probably could not resurrect the figures of what it cost. But I would gladly do it all over again. The men were worth the effort. They were responsive and responsible. They had gone through a turmoil which could be viewed in the worst light as almost a mutiny, and in the best light as an effort worthy of a Patrick Henry on behalf of the traveling public and the pilots who fly them.

The old familiarity that used to fairly crackle over the airwaves when I checked in with the tower from number 808 LJ has long diminished. Though I remember with particular fondness the number of controllers who had told me, off-the-record, that they gladly would have given me equal time with Air Force One, I no longer make it my business to let controllers know the side number of my aircraft. I'm simply content to accept their excellent service. But I will never forget them and the valiant fight they waged.

I would be remiss if I failed to mention that in the middle of one of the most unpleasant, in fact miserable, periods of my life, the air traffic controllers (long estranged in the sense of day-to-day contact) reappeared in the spirit of simple friendship.

In December 1973 when my own trial* in Jacksonville, Florida, was bogging down to the point where it was nothing more than a dirty debacle and, in my opinion, a low point of jurisprudence, justice, judicial *and* prosecutorial conduct

---

*In May 1973 I was indicted for mail fraud, along with my former client Glenn W. Turner and eight of his business associates. Twenty-seven months later a federal judge dismissed the charges against me without my case ever having gone to the jury, even though I had been on trial for seven months in Jacksonville. For all the gory details, see Chapter six of *For the Defense*, (NAL, June 1976).

(and almost everything else I can attribute to a sad situation) the boys from PATCO came forward.

They said, "We're having a PATCO party in Jacksonville, and we'd like you to attend. We've got a special table set aisde for 'PATCO has-beens.'"

I went to that party with my wife Lynda, enjoyed every single minute of it, and seated at the table for PATCO has-beens were all the boys who started the organization. Not the new leaders, not the present hierarchy, but the ones who were in on the ground floor and remembered those beginning days when an idea became a movement became a phenomenon—one which troubled and indeed at one point stopped the government of the United States, but only long enough to make it reconsider its standards of safety.

As PATCO has-beens we had—under the circumstances—the best Christmas party of 1973.

5

Natural Factors

I've had my share of white-knuckle flights where the sudden appearance of bad weather caused the pucker factor to increase dramatically. But one of the most chilling tales I've ever heard about the potential danger of weather conditions is told by Dave Savage, my chief pilot.

Several years back when he was flying for Executive Airlines, a large commuter service catering to businesses along the East Coast, Dave was in command of a 19-passenger DeHaviland Twin Otter. It was late October, a cold day with just a little bit of fog, and Dave was the captain on a flight from Portland, Maine, to Boston, with a crew of two and fourteen passengers.

Ordinarily, the trip takes forty minutes. That day, as it turned out, it took almost four times as long—and it could well have ended in disaster if the pilot had been someone less skilled and cautious than Mr. Savage.

Before takeoff, Dave made his usual check of the weather conditions at his destination and was told that Boston had some fog but, as he put it, "nothing drastic." The flight plan called for him to fly from Portland to the Kennebunk VOR navigation point, then on the Downey Airway to the Ipswich Intersection, from which point he could proceed directly into Boston.

His plane carried enough fuel for it to fly at the usual cruise speed for two-and-a-half hours. Under normal conditions that's enough fuel to get there and back a couple of times. However, as he soon found out, conditions were anything but normal.

When he reached Kennebunk, only twenty-four miles from Portland, air traffic control put him in a holding pattern and, surprised, he called in.

"What's going on?"

"Boston just started getting some fog," the controller told him, "and they're starting to get some delays en route."

He only had to stay in the holding pattern for a single turn and expected that the thirty minutes from Kennebunk to Boston would be routine.

When he neared his next checkpoint, Portsmouth, New Hampshire, he was again directed to hold. This time he decided that he would take part of his hold en route, which he did by simply slowing down. Good careful pilots will do this, rather than fly at normal speed to the hold point and then go around the required number of times, because it saves fuel. He only had to make one hold at Portsmouth.

When he got to Ipswich, he soon realized that there was reason for some concern: he had to go around the holding pattern nine times.

At this point he had been in the air for an hour. By flying at the reduced-power setting he had just under two hours of fuel left, so he was concerned but not worried.

When he finally got near Boston he was put in still another holding pattern, and he found out that the only reason he'd gotten so close so soon was that other airplanes in the pattern were leaving and going on to land at alternate airports.

Executive Airlines was located at Boston's Logan Airport, and Dave called in on the company frequency to see what he could learn about the extent of the weather conditions. He was told that the airport was blanketed with almost solid fog, but that every half hour or so it let up enough so that a single plane could get in. The visibility was down to a quarter mile. Now Dave began to sweat just a little bit.

He also learned that no one had been able to shoot an approach in the last half hour. By this time it was 4 o'clock in the afternoon, and in addition to the fog, it had begun to drizzle. He started to think about going on to Bradley Field in nearby Connecticut.

To choose the alternate would mean, he figured, that he'd have to leave the holding pattern near Boston with one thousand pounds of fuel so that he could get to Bradley with forty-five minutes of fuel in his tanks. He decided to go to Bradley.

He checked with the company again and learned that almost the entire East Coast was closing down. Boston was down. New York was down. One of the company's planes had just made the last landing at Providence before it closed, and Hyannis was also down.

The closing of both New York and Boston meant that a lot of planes were suddenly heading in the same direction. When Dave got to Bradley he was told that he was number fourteen in the holding pattern!

But let him tell the rest of the story:

"By the time I was approaching the final approach fix, after waiting my turn, the visibility was down to about a half mile. I checked the weather again, just as an aircraft in front of me passed the final approach fix. Now, if you haven't crossed the final approach fix and the weather drops below the minimum for approach visibility, you're not allowed, legally, to execute the approach. If you are past the final approach fix, you can come down, take a look, and if you can see it, you can land. If you don't see it, naturally, you can't land.

"But at this point I am down to forty minutes of fuel left. I'm five minutes under my legal minimum—because I've been vectored all around the world, because of the fourteen airplanes in front of me. But where can you go in forty minutes? And this isn't forty minutes of nice flying, looking around for someplace safe to go in. This is forty minutes until the engines quit. And the plane stops flying.

"So I pass the final approach fix, and I'm going on down and in. I tell my copilot, who is looking a little nervous at this point, 'We're just going to keep on going down the glide slope and the localizer [two electronic aides that "tell" the pilot, through his instruments, how to approach the field]. I've flown this approach a couple times here before, and they've got a good glide slope and a good localizer.'

"So I'm just going to keep everything crossed, meaning the glide slope and the localizer, just keep them in the center. And I kept them in the center so well that I was proud of myself.

"When I got down to decision height, if the tower had said something like, you'll have to go around again, I was at the point where I would have had to declare an emergency and keep right on going in, because I hadn't enough fuel left to go anywhere else.

"So, I got down to my minimums, glanced up for a second, and couldn't see a damn thing.

"I got down to about 40 or 50 feet, still using just the glide slope and the localizer, and I glanced up and saw one white light. And then another white light. These were the center lights on the runway.

"I didn't see the rabbit as I went over it. The fog was too thick.

"As I went over these two white lights, I thought, 'Well, it's one of two things: either I've got the center runway lights, or I've got the lights on one side of the runway or the other. Under these conditions, I got nothing to lose. Even if I touch down on one side of the runway, the most I can do is knock out a runway light or something like that. If I go around, and *don't* get to shoot another approach right away, I'm in deep trouble, and the worst thing that can happen is to have an uncontrolled crash when both engines flame out, and go in someplace when you can't see the ground. At least I know where I am.'

"It was approximately five seconds after I saw the first light that I raised the nose and the wheels touched. Maybe less than five seconds, maybe two or three seconds.

"I looked out and I couldn't see either side of the runway. I had to ask the tower how far down the runway the first turnoff was. It took me five minutes to taxi in to the gate.

"Later, I learned that two aircraft in front of me had missed the approach and had to go on someplace else. And everyone behind me missed the approach. No one else got in that night."

* * *

Dave Savage's experience indicates several important truths, not the least of which is that his passengers were very fortunate to be in the hands of such a skilled pilot. It also indicates that sudden changes in the weather can cause serious problems in the air. In fact, weather is one area of flight safety where increased human vigilance—and experience—can pay instant dividends.

Pilots who fail to pay adequate attention to weather conditions do so at their own peril—and that of anyone else flying with them. A few years ago, a pilot on a flight very similar to that just mentioned found himself in much the same situation. But he didn't take the precautions that Dave did, like cutting down on his speed to conserve fuel, and when he missed his approach he elected to go on to another airport. On the way, he passed several other fields where he could have landed easily, but he was reluctant to declare an emergency. Instead, he ran out of fuel over Long Island Sound. The plane went in and everyone on board was killed.

If there is one weather-related crash that is uppermost in the public mind, it is that of the Eastern Airlines Boeing 727 that crashed while trying to land at Kennedy International Airport on June 24, 1975. That was, for lack of a better term, the now infamous "wind shear" crash.

At 4 o'clock in the afternoon, an Eastern jumbo jet, a Lockheed L-1011, experienced such a vicious downward pull to the right as it was about to land that it had to apply takeoff power. The pilot decided not to try another approach at Kennedy and headed for Newark, where he eventually landed without incident, but on the way back up he radioed the controller at Kennedy and reported,

"We had a pretty good shear pulling us to the right and down. . . . We were on course and, ah, down to about two hundred and fifty feet. The airspeed dropped to about ten knots below the bug [a warning device the pilot sets to let himself know if his landing speed drops too low]. And our rate of descent was up to fifteen hundred feet a minute. So we put takeoff power on, and we went around in a hundred feet."

The next plane to attempt a landing made it, but just barely, and its pilot was sufficiently shook, and professional, to contact the tower and warn the controller about the wind conditions. The pilot, who had just landed his DC-8 cargo plane (owned by the Flying Tigers line) that also carried eight passengers, said, "I just thoroughly recommend that you change the runways and land northwest. You have such a tremendous wind shear down near the ground on final."

Apparently, the controller was worried that if he changed runways he would be sending aircraft into another hazard; there was a severe crosswind buffeting the alternate approach. He replied, "Okay. We're indicating wind right down the runway at 15 knots when you landed."

This did not appease the Flying Tiger pilot: "I don't care

what you're indicating. I'm just telling you that there's such a wind shear on the final on that runway you should change it to the northwest."

Several miles off in the distance, the pilot of another Eastern plane, the Boeing 727 designated Flight # 66, apparently was listening to at least some of the conversations. When he received and accepted his final clearance, he told the controller, "Okay. We'll let you know about the conditions."

He never was able to. At five minutes after four, in the process of wrestling with a vicious wind shear—a condition in which adjacent winds of extremely high velocity blow in opposite directions—the plane struck several approach light stanchions about a half mile from the edge of the runway.

Like a gigantic metal bird wounded and trying to escape, the plane rose for several moments. But it hit more stanchions, turned completely over, and burst apart in flames and explosions.

The eventual death toll was 112. Somehow, twelve people survived, doomed to be always reminded of their "good luck."

And the public had a new term—wind shear—to add to its lexicon of anxiety.

The immediate debate had to do with the question: Did the weather cause the crash? It was not an easy question to answer, because two planes landed right after the crash (one a Finnish jetliner and the other a small prop-driven private plane, a Cessna 180), and because neither the pilot nor the copilot had survived.

But one thing was clear, the vicious wind conditions were certainly a strong contributing factor. And we, pilots and passengers alike, were warned once again that severe weather conditions can be every bit as deadly as human error.

* * *

It's been my experience, as both pilot and lawyer, that most accidents are associated with weather, as opposed to, say, mechanical failure. The final determination may often be pilot error, but the chances are great that the error would not have been made in good weather. Thus any discussion of air safety must concern itself with the topic of weather.

Most often, the man on the ground thinks about thunderstorms, hurricanes, and other examples of violent weather conditions in relation to air safety. But there are many other conditions, seemingly mundane, that can be vital. For example, consider the simple element known as ice.

Even the most experienced pilots shudder reflexively at the thought of what ice can do to an airplane. On the ground, ice on the wings means no takeoff, at least not until all of it is removed. In the air, fast-forming ice can turn a flying machine into a stone.

Before you start to cancel your winter trip to Florida, let me explain that ice is rarely if ever a problem for the major airliners. In fact, I doubt if ice bothers them one bit, because they have all sorts of sophisticated equipment to prevent it, or to blow it off as soon as it forms.

One device is an anti-icer, which must be turned on before the ice accumulates (remember that the cockpit recording of the last minutes of TWA 514's flight mentioned checking to see that the anti-icers were on), and the other is called a deicer, which works after the ice has purposely been allowed to accumulate.

Propellers now have electric pads on the outside of the blade closest to the hub, and they keep the temperature too high for any ice to form. In earlier planes they used to spray alcohol to do the same thing.

Today's jet engines take "bleed air," which is simply very hot air straight from the compressor, and pipe it around. They pipe it to the front of the wings to keep them hot. But the big black "boots" that passengers can see on the wings and tail surfaces are the oldest form of deicing equipment. The ice is allowed to build up, intentionally, to about a half an inch or an inch in thickness, and then the pilot hits the button. The boots expand as compressed air is shot in and the ice pops off. And in certain icy conditions you just have to keep doing that to get rid of the ice.

Hot wings are much better, but they take a lot of power off the airplane. With an anti-icing device you don't have to worry, unless of course it stops working!

Sometimes, if we (in my Rockwell Commander) get some ice accumulating around the inlets of the turbines, we have to throw some heat at it and then be ready to use the igniters in case the fire goes out. If you melt a big bunch of ice and it goes into the turbine it can blow the fire out, but if you have the spark continually igniting, then you don't have any problem. And if you anticipate the ice, that's even better.

The worst kind of ice to run into is heavy clear ice (as opposed to "rime ice," which is slow to accumulate and rather closely resembles the stuff that forms in your refrigerator when it hasn't been defrosted in a while) that is just like ice cubes.

There was a report a few years back from a TWA Constellation that it was picking up clear ice at the rate of *three inches a minute*.

What makes heavy clear ice so dangerous is that if you have a malfunction in your boots so that they won't pop, you will pretty soon cease to be a flying machine. This happens because clear ice will form over everything, knock out the radios, increase the weight of the plane tremendously, and actually increase the stalling speed so that you get to the point where the engines won't keep it in the air anymore unless you're descending.

Dave Savage had a horrible experience once with clear ice. He was flying at 3,000 feet on his way into Augusta, Maine, and he had to descend through the clouds. Ice began to build up at such an alarming rate that he went back up again. His only recourse was to make his approach at a much faster rate of speed than normal, which never pleases the passengers. He did so, but by the time he broke out of the clouds he was so heavy with ice that at 500 feet above the ground the airplane—carrying more than a dozen people—simply stopped flying. It was the first time in more than twenty years of flying that he'd ever had an airplane quit flying on him in the air.

He glanced at his rate-of-speed indicator and saw that it was in excess of 4,000 feet a minute. It took almost his entire bag of tricks, but he got the plane to the point where he landed at the rate of 1,000 feet a minute—which is about five times greater than the normal touchdown speed.

It was, to put it mildly, a hard landing. And in a hard landing the airplane usually bounces, but this one was so heavy with ice that it didn't bounce at all. As Dave said, "The co-pilot and I looked at each other, we parked that airplane, and we made the rest of the passengers go on to Waterville, Maine, by automobile. Scared the hell out of me."

Fog, as I mentioned earlier, can be a real problem and obviously rain can cause difficulties, especially on landing because an accumulation can create an effect known as hydroplaning (the thin cover of water prevents the tires from making contact with the ground) which means you won't stop in the normal amount of time. If something is in your way—like a building or a fence—you can have made the best landing in the world and still be in deep trouble.

The most serious of all weather conditions, though, is undoubtedly the thunderstorm. Nonpilots probably think that it would be more dangerous to fly through a hurricane, but actually the reverse is true. Certainly no one should do it as a lark, but the National Weather Service airplanes and those of the military routinely fly through hurricanes on patrol. Of course, these are especially rugged and very well-equipped planes, and the patrols are often somewhat dangerous. But the idea of it being impossible to fly through a hurricane is a misconception.

A thunderstorm is—to mix metaphors—a wind of a different color. No pilot in his right mind intentionally flies through the center of a thunderstorm. No one has any accurate statistics on how many planes have simply disappeared because they went head on into a thunderstorm. Later, all that turns up is scattered pieces of the wreckage—and there have even been instances where pieces of the airplane were found scattered miles from each other.

A thunderstorm is so dangerous because it contains all of the worst surprises that weather can hand out. It has turbulence so severe that it can rip an airplane apart literally: the wings are bent so often in opposite directions that they eventually snap off just like a coathanger that has been bent back and forth.

The wind shear that caught the Eastern jet at Kennedy was an almost freakish occurrence, but a bad thunderstorm has constant wind shears. Moderate turbulence is something most airline passengers have experienced and it may discomfort them somewhat, but it usually results only in the plane's speed fluctuating and the people inside bouncing around from time to time. Severe turbulence is something else; the official definition of severe turbulence is, "when things start flying around inside the airplane." And that's no fun at all.

Not only do people lose meals, present and past, but there have been cases where lives were lost because someone was in the aisle or sitting unstrapped when turbulence caused the plane to drop or rise several thousand feet in seconds.

There have been two cases in my experience where military pilots had to bail out in thunderstorms. One poor soul bailed out into the vertical air currents typical of thunderstorms, and instead of going down he went *up*. It took him twenty-five minutes to get to the ground from 30,000 feet. By the time he finally made it, he was black and blue from the wind smashing him continually.

Not only do you get ice in a thunderstorm, but you can get hailstones as big as your hand, which are extremely hazardous. A thunderstorm also may contain a tornado within it. In fact, there are two types of thunderstorms—those that have tornados and those that don't. Another problem with thunderstorms is that their size is not necessarily related to their intensity. We flew into one once on the way back to Atlanta from Nassau in my Beechcraft Duke, and although the storm was small, it was quite fierce. You know the turbulence has been bad when your seat belt hurts you because you've been slammed against it repeatedly. That trip contained some of the worst weather I've ever flown through.

Fortunately, modern airplanes are far better protected against thunderstorms than those of a few years ago. Perhaps the single most effective tool is the weather radar. All the commercial airliners are required to have weather radar, as are certain other types of planes, and many others (like mine) that are not required to have it do so because it is so helpful.

This type of radar paints a thunderstorm beautifully. It shows you where the center of it is, the next heaviest area of precipitation, and where the light precipitation is. You can usually fly through the light area and bounce around some, which is slightly uncomfortable but not very dangerous.

The area of heavy turbulence is normally the area with the heaviest rain, and what the radar picks up is this heavy rain; it shows up on the scope as a dark spot, and that's the area a pilot wants to avoid at all costs.

There are occasions when a pilot inadvertently finds himself in this "dark spot" area. It happened to me once and I could barely control the airplane. I made a 180-degree turn within about five seconds of getting into it, but it took me quite a while to get out of it because of the way it threw me around. When I tried to make a left turn it would roll me over on my back, and then turn me in the opposite direction. My four passengers were less than delighted when we shot up 2,000 feet in five seconds.

The curious thing is that when I got out of it, the controller told me that another plane had gone through only three minutes before and reported only light to moderate turbulence. Yet, it had been definitely severe when I went in. In fact, it had done some damage on the ground.

Thunderstorms are heaviest in the summertime, but you can get them any time of the year. When there's a lot of instability and a lot of sharp differences in barometric pressure, there will be rapidly moving systems that are likely to have thunderstorms. And they can come steaming across the ground at fifty miles an hour.

The most dangerous kinds of thunderstorms are those that form in lines, which means you can't really get around them. There'll be a solid line of thunderstorms, kind of all joined together, and it will be a whopper; unlike scattered thunderstorms which occur on a summer day and which you can get around (unless they're embedded in some other form of cloud.) Normally, though, a pilot can get fairly close to a thunderstorm without any danger.

Everyone has seen the beautiful long, billowing clouds that may look like an anvil at the top. They go way up and look wondrously peaceful. But they can be quite dangerous because they are building up a great deal of power in a short time, well before the rain comes, in many instances. It is not uncommon for a pilot to be climbing at 4,000 feet a minute and notice that the top of the cloud is going faster than he is.

What makes the thunderstorm so dangerous is the intensity, the frequency, and the duration of the turbulence. The definition of turbulence is simply a sudden change of direction in the air. And normally that's no problem; you go over one bump, and that's it. You settle down. But a thunderstorm is made up of a whole bunch of bumps, so you go slamming through one, up one and down another, and then still another. Not much good for the nerves, and after a while hard on the airplane.

Fortunately, airplane construction is so advanced that a plane can take a very great deal of this kind of stress. There have been cases of structural failure due to thunderstorms, but they are so few as to be almost rare. The serious case is when a thunderstorm aggravates other problems.

This discussion of thunderstorms may help the reader to understand why all pilots are so concerned with the weather. Those who fly the big jetliners have a slight edge on the rest of aviation in that all the major airlines have their own weather forecasting and reporting services, and there are times when their forecasts are more accurate than those of the National Weather Service. The passenger can rest assured that any decision affecting the safety of a flight will be based on solid information.

When problems arise, it is usually because the weather has changed suddenly. One should remember that the weather calculations and measurements are taken about ten to fifteen minutes before the hour, so that the report given some five or ten minutes after the hour indicates what the weather conditions were *then*—and not what they are at the moment the pilot calls in. When you take into consideration the average length of flights, it should surprise no one that planes occasionally run into weather-related problems.

Unlike the airline captains, all other pilots have to depend on a variety of sources for their weather information. There are a few independent services, but they are rather expensive for the smaller noncorporate operators. Most pilots get in the habit of turning on the *Today* show every morning, but not simply to catch up on the news; they want to watch the large weather map in the background because it shows the general weather picture throughout the entire United States.

When someone like Dave Savage or Dick Bedell is scheduled to fly with me, Lynda, and perhaps one or two other people from Boston to, say, the Enstrom factory in Menominee, Michigan, he will start checking the weather the night before. Then prior to takeoff he will check it again, always trying to determine what the trends are—because, as I said above, it often changes and it is not of-the-moment information. He must learn that weather conditions are within the FAA minimums that govern our type of aircraft for the particular destination, and he must select an alternate site in case things change en route. And the minimums for the alternate have to be higher than for the destination; in other words, if he has to divert, he must do so to a field with better weather conditions, not the same or worse. All of this, of course, is in the interests of safety.

Then when he's up he continues to check or monitor the weather—winds aloft, ice, turbulence, and so on—to see how the estimates compare with the actual weather, again with an eye toward greater accuracy. Most of the time the actual weather is what the National Weather Service says it will be.

Then as he gets closer to Menominee, he checks the destination weather once again. If it is good, then he usually doesn't check it until just before we arrive, but if it is marginal, he will probably check it every hour. Is it steadily getting better? Worse? Staying the same? All of this information is cranked in to his final decision as to whether or not we can land when we get there.

With the possible exception of fog, foul-weather flying—and landing—is not the problem it used to be, for the simple reason that we now have such sophisticated equipment, both in the airplane and on the ground, for instrument flying. For an experienced instrument pilot, a night landing in the soup at a modern airport equipped with an Instrument Landing System (ILS) is almost routine. But change the airport conditions and the pilot's experience and you have a whole new ball game.

(TWA 514 was making a *nonprecision* approach, which means the runway it was heading for did not have an electronic glide path for him to follow, and he was in extremely adverse weather conditions.)

Finally, to conclude our trip to Menominee, Dave or Dick would have to be concerned with what the weather has done to the airport itself. If it is raining, it will not only cut down on his visibility, it will make the runway slick. If a snowstorm has hit Menominee (hardly a rare event in the Upper Peninsula) snow or ice on the field will be a problem. Sand helps to add more friction, but if the field dries and the sand remains (or at a different time of the year if high winds have blown sand and dirt onto the field) it makes it *more* slippery.

Another factor he has to consider is the wind. Pilots prefer to land into the wind, or as close to into the wind as possible, because it cuts down their speed in relation to the ground. If he can't land into the wind, he may have to land with a tail wind, and although no one prefers to do it that way, most airplanes have about a 10-knot tail-wind-component allowance. It is not at all unusual for a pilot making an approach with a tail wind to go back up in the hope that the wind will change direction and he can land into it.

So as long as the turbulence isn't too bad, and the field

isn't socked in with fog, and nothing goes wrong with the airplane, we land at Menominee with no problem whatsoever—which is exactly what we do more than 99 percent of the time.

After all this talk of weather, the lay reader might be interested in knowing just how the weather measurements are taken.

There has been some criticism in recent years that the men and women who measure the weather at certain airports are not properly trained—one gets the picture of someone sticking his head out the window and squinting at the darkening sky—but few pilots I've talked to agree with this. The training is provided by the National Weather Service and it's a good course.

There are several ways of measuring cloud height. One is almost poetic in a sense. A black balloon with a weight attached is filled with helium until the point where it lifts the weight off the table. Then the weight is removed and the balloon is tied. Now it has a known lifting capacity which enables one to determine the rate at which it will climb. Once outside, the balloon is released and a stopwatch is used to see how long it takes to disappear. The balloon man then consults his table, which tells him how high the clouds must be if it took X number of seconds for the balloon to disappear. This measurement is called, not too surprisingly, the balloon ceiling.

Another method of determining cloud height is called the measured ceiling, and there are two ways of doing it. One is to use a light shining straight up (and obviously this method is preferable at night, but it can also be used during the day because the light is of a great intensity) until it hits the clouds, and then a button is squeezed, which locks in a device known as an inclinometer, or angle meter. The next step is to read the final height by means of trigonometry, using the ground and the angle and the cloud ceiling as the reference points. The second method of getting a measured ceiling is to use an automatic device equipped with a light sensor that detects where the light is hitting the clouds and gives an automatic reading.

Oddly enough, there are places where actual human beings go out, look up at the sky, and make an educated guess. And they are surprisingly accurate. Finally, pilots often provide information about cloud height, because when they come out of the clouds on an approach they have only to look at the altimeter to see at what height they left the clouds behind.

Visibility is another bit of information that the pilot wants to know. That is also measured in several ways, one of which is simply by looking out toward a prominent landmark, such as a tower or a church steeple, that is a known distance away. If it is a mile away and they can just make it out, they give you that distance or perhaps a little less. If they can see it clearly but not the next point a mile farther out, they'll give you their best estimate on the difference. This is the eyeballing method, and it can be quite accurate. Another method is to use the Runway Visual Range system at the end of the runway, if the airport is equipped with one, and determine how far the system's light sensor can "see." Because the light is always of a fixed intensity, when something (clouds, fog, rain, snow, blowing sand, smoke haze, whatever) gets in the way and cuts down the intensity, this is reflected in the measurement given. And that's fairly accurate.

An airport gives its average visibility and frequently qualifies this by direction; it might be two miles to the south but only a half mile to the west because of factory smoke or a fog bank.

Such services also provide "forecast temperatures aloft," which can be very important to a pilot who does not have the latest in de- or anti-icing equipment.

Modern aircraft are equipped to handle almost any weather condition except severe turbulence and the thunderstorms that usually accompany it, but from time to time there arise combinations of weather hazards that stack the odds rather heavily against even a well-qualified pilot. One good example is the combination of an air inversion on a cold day (when you descend toward the airport and the air suddenly turns colder, you can pick up ice at a very rapid rate) and very low ceilings and visibility, plus occasional crosswinds. Normally, when an inversion occurs, the air is comparatively stable and strong winds will not be a factor—but there are some frightening exceptions to this general rule.

The combination of heavy icing on approach (when power is low and the anti-icing equipment is at its least effective point) together with a coating of ice on the runway (which will reduce braking power to zero) has the effect of closing the airport to all aircraft that do not have decelerating devices, such as thrust reversers on jets or reversible props on propellor-driven planes. The possibility of breaking out at the last minute, or missing the approach and having to climb up once again through the clouds while burdened with a heavy coat of ice, can lead the wise pilot to abandon the effort before it is begun.

Perhaps the most important moment in any flight occurs when the pilot analyzes the combination of weather factors awaiting him, which he can perceive from a variety of sources (the weather service, approach control, the tower, and the conversations of other pilots which he can listen to as they negotiate the final phase of flight prior to touchdown), and then makes an appropriate decision. At least one hopes it is appropriate!

If there is plenty of fuel left for an attempt at an approach and landing and the aircraft is not so heavy that it can't carry a fair load of ice through that landing procedure and a subsequent missed-approach procedure and all of the equipment is operating properly and the crew has had plenty of recent training in marginal weather flying, an effort to shoot the approach should not be considered unduly bold. (That sounds like a lot of "ifs," but it really isn't.) By these standards—and because of the way they plan and program their flights—one would expect most airline pilots to make at least one pass at the field, unless the conditions completely preclude minimal safety conditions.

The less current and disciplined pilot, whose equipment might not be so sophisticated, might well be able to handle a *single* weather obstacle; but he would be ill-advised to take on a combination, especially if he lacks any of the items mentioned above.

Unfortunately, a great many of the nonairline accidents (and too many of the airline accidents) result from the failure to resist just such a challenge.

The FAA runs the Flight Service Stations and the National Weather Service operates weather stations, and the cautious pilot makes use of both of them. When neither is available, he turns to his old friend the telephone.

Which leads rather neatly into my final point. Weather information represents one of the few areas in which the average airline passenger can do something about his or her own safety.

CLEARED FOR THE APPROACH

Most nonairline pilots use the National Weather Service to get their information, and they get it over the phone. There is no reason why a worried traveler who is about to entrust his life to God and his copilot in the form of a major airline cannot do just what the general aviation pilot so often does and pick up the telephone and call the National Weather Service, which is listed in all major city phone books, and find out what the weather is predicted to be in Cleveland or Boston or Houston or Sacramento on the day or night when you are scheduled to arrive.

Most pilots are asked to give the side number of their plane, but that is because the government employee on the other end gets brownie points depending on the number of calls he handles, and I see nothing wrong with simply saying, "I am to be a passenger on flight number so-and-so on such and such airline." I doubt very much if the person on the other end will refuse the information.

There is a slight danger in this, however, in that pilots already have to hang on the line a long time before it is answered, but if a concerned citizen calls only when he or she is truly concerned, it should not overload the system unduly.

Let's face it, there are times when nervous people will be a lot better off if they simply postpone a trip. Only one thing is certain about the weather, and that is that it will change. Just ask any pilot.

* * *

At the outset of this book, I said that with the exception of acts of God and nature (which many people feel are synonymous) there is no such thing as an unavoidable air crash. If you remain unconvinced by this point, it may be because you have a vivid memory of certain disasters related to air travel. "If it's all as safe as he says it is," you might be thinking, "then what about near-misses, and why doesn't he mention anything about the birds, or acts of terrorism?"

Fair question, up to a point. It's my opinion that in most of these cases, the risk factor has been exaggerated way out of proportion to the actual danger involved, at least insofar as the average airline passenger is concerned. And that's why I prefer to deal with these happenings and phenomena in one section. The media have already done a pretty good job of

FEAR OF FLYING

scaring people, and I don't want to add to the problem. I'd like to try to set the record straight.

Let's begin with a subject that has poetic, perhaps also philosophical, overtones—the birds. It has to be one of the greatest of ironies that the noble bird, for so long the envy of earthbound man and the symbol of flight itself, should pose a threat to a modern airliner. Depending on your point of view, there is something either inherently amusing or infuriating in the unusual (but not rare) standoff between a gigantic jet and a flock of scrawny birds: while the airliner filled with people waits at the edge of the runway, its thirsty engines gulping fuel even while idling, a man with a shotgun or some other noise making device has to drive to the end of the runway and scare away a bunch of sea gulls or starlings. Such occurrences give nightmares to the airlines' efficiency experts, and they puzzle tired, anxious-to-get-home passengers who see that the field is clear of traffic and can't understand why the pilot doesn't take off.

But, as the pilot well knows, the presence of those simple birds presents a very real threat to the lives and safety of everyone on the airplane.

Some history is in order. In the early days of flight, in fact up to the era of the jet airplane, birds were not a serious problem—unless a Waldo Pepper type broke down his windshield to feel the wind in his face and got a good-sized bird instead. Those who know their physics will remember that given sufficient velocity even a piece of straw could be driven through a steel girder. Thus it should not surprise people that a small bird flying at say, twenty miles an hour, can be a hazardous object for a plane streaming along at ten times that speed.

As a result of this potential danger, aviation manufacturers devised what was known as the bird-proof test, whereby a chicken (I kid you not) was fired out of a cannon-like gun and aimed directly at a windshield to see if the reinforced glass could withstand the impact. It should go without saying that this was a messy business and not the favorite assignment of those who worked in the factories where such tests were run. In the early days of jets, chickens were shot into the whirling turbine; if the chicken passed, so did the engine.

There was a well-publicized incident in the late fifties where a sea gull—perhaps a soul brother of Jonathan Livingston—struck a DC-3. The bird went through the leading edge of the wing and finally came to permanent rest inside the fuel tank at the rear of the wing. Such was the force of the collision between these two disparately-sized flying objects.

And even though the windshields of most airplanes were finally made strong enough to resist breaking or shattering, the problem was still not necessarily solved. Dave Savage, my chief pilot, tells of the time he saw the results of a meeting between another sea gull and a military training jet, an F-9. The bird struck the windshield (or vice versa), and when that part of the airplane did not give, the seagull's body was driven upward against the face of the glass until it smashed into the top of the Plexiglas canopy, whereupon it burst through, splattering plane and pilot with its own remains.

For some reason, one result of such an accident is an almost overpowering putrefaction. In other words, what's left of the bird stinks like hell. As Dave related it, it was several days before even his friends would sit down in the near vicinity of the pilot (who had managed to land safely) and several weeks before all olfactory traces of the incident left the airplane.

But I don't mean to make light of what was to become, with the introduction of the jet, a serious problem, one that continues to the present day.

A good-sized jet airplane standing at the edge of the runway, its engines cranked way up in anticipation of the great power needed for takeoff is in effect a giant vacuum cleaner. It will suck up anything in its path. Which is just what happened on October 4, 1960, at Logan Airport in Boston, not far from my home.

Just as an Eastern Airlines Electra was taking off, a flock of starlings reeled about in total confusion. A number of birds were sucked into the pair of engines on the plane's left side. Immediately, the pilot pulled back the power on the opposite wing, but as the birds were digested and whirled through the huge Mixmasters on the port side, those engines returned to full takeoff power. The resulting imbalance caused the plane to roll violently to starboard, driving it down and into the bay. Of the seventy-two people on board, all but ten lost their lives.

The machine in the garden had produced another tragedy. Fortunately, that bird-related death toll has not been exceeded to date. But as recently as November 1975, a flock of sea gulls was sucked into the engines of a jet airliner taking off from New York's Kennedy Airport. The pilot crash-landed, and all of the people on board—139—were safe. No one was seriously injured, but the plane was burned beyond repair. Thus, it seemed to many people that something must be wrong if, in this age of technological advances that would have seemed miracles a few decades ago, a bunch of birds could fell a gigantic flying machine. But they can, and do.

As surprising as it may seem, no one has come up with an effective way of keeping birds off the runways—save going down there and scaring them away long enough for the plane to take off.

Immediately after the 1960 Logan Crash, people became concerned for the first time with the problem of birds knocking out a plane's power plants, and a variety of programs were begun to try and remedy the situation. Logan started using the Massachusetts state police, who took their shotgun practice at the airport, to scare the birds away. Other airports, among them Washington's National, experimented with recordings that carried the *sound* of shotgun blasts, and several others actually put up scarecrows. But, to the great dismay of the airport officials, after a few days the birds began to sit on the scarecrows. A few years ago, National installed foghorn-type noisemakers; today the birds sit on the large horns and when they sound, the birds rise up a few feet, flutter slightly, and settle right down again as soon as the noise is over.

There just isn't much else you can do. The average passenger, however, should know that no captain in his right mind will take off until someone or something has convinced the birds to leave the runway, at least temporarily.

The problem of bird ingestion, as it is properly called, is actually one segment of a larger problem known as FOD, or foreign object damage. Just as a bird or birds can be swept into the maw of a hungry jet engine, so can almost anything else of insufficient weight that finds itself in the path of a large jet revving up for takeoff.

The most serious personal experience I ever had with the overall problem was caused by an object that, after the fright and the flight were over, turned out to be a badly chewed-up piece of orangewood.

I was flying a borrowed Lear Jet, owned by a client who faced the possibility of many years in prison, and after conferring on his behalf in Pennsylvania, had to be in Baltimore, Maryland, for an appellate argument for another client. One does not come late to appellate arguments. I took off on a very windy day (the winds were gusting up to forty miles an hour as a thunderstorm passed over the field) when, out of the corner of my eye, I noticed something fly up off the runway.

It flew right into the main turbine, and the dials began to sing a sad song. I immediately pulled the power back to 80 percent, because the compressor blades were bent (which means you can't pull full power) and proceeded on to Baltimore at about 490 miles per hour, on one engine.

This happened during the Medina case, when all the hours were precious and I had no intention of missing an obligation unless there was nothing I could do about it. Fortunately, the Lear Jet had so much power that I really didn't have to worry. But when I got on the ground I called my client and told him to come and get his airplane. I found out later that he not only sold it "as is," but he made $13,000 on the transaction.

To make the point that it is not just birds that cause FOD damage, consider the problem of the two TWA 1011's that were sent out from Boston on the same day not too long ago. In fact, it was New Year's Eve. They had not been fully deiced on the top of the fuselage. Both lost the center, or number two, engine on takeoff, from ice going back through the intakes (which are in the vertical fin). Both went down, and although neither one was very serious from the standpoint of in-flight emergencies, both had to have new engines installed. The point is that whether you are talking about birds or ice or pieces of debris—anything other than air that passes through a jet engine is a foreign object.

Before certain readers panic, however, it should be stressed that turbine engines in general are far more reliable than the conventional older piston-type engines. There are fewer moving parts and they are not fighting one another like a piston engine, with the pistons slamming up and down. They hold together pretty well, and they tend to go for a long time without needing an overhaul. Believe me, I wouldn't rely on them the way that I do if they were not a decided advancement in the state of the art of flying.

But they *are* sensitive to good-sized objects that get caught inside.

It should be understood that a jet engine will normally chew up anything flying into its "mouth" unless that object is fairly solid. There is another problem, however, and it brings us right back to the birds. A single bird—starling, sea gull, whatever—will pass right through the engine; but if you run into a flock of them, you face a different problem, and that is that the number of birds (or other foreign objects) will cut down the air that is essential to the engine's operation.

Most birds, unless they are unusually large, will pass through the engine without doing any damage at all. But if there are enough of them they will cut down the air intake, thus crippling the efficiency of the turbine.

For example, the birds that downed the Eastern Electra back in 1960 were not large birds, but there were enough of them so that they blocked off the air. (Many times a pilot will not even know that one of his engines has "digested" a bird.) The proof of this is that as soon as the birds had passed through the engine, the turbines returned to full power.

(It is interesting, and perhaps instructive, to note that in a recent Sunday magazine supplement article that discussed the problem of bird ingestion, it said that the plane that crashed at Logan in 1960 was felled by the loss of one engine. That is simply not true. An Electra can lose one engine—for any reason—and still remain airborne and perfectly safe, using its three other engines.

As Dave Savage puts it, "That plane can fly on *two* engines all day long. As a matter of fact, when Navy fliers go out on patrol in four-engine planes, they often shut down the two outboard engines to conserve fuel and just fly on the two inboards. If they have a problem, if they lose either one of the inboards, they have plenty of time to start up one of the other engines—whichever one they want.")

There is one final calmative point that the average air traveler should keep in mind, and that is that every major airport authority knows that if he doesn't do something about the presence of birds (or any other foreign object) on or near the runway, his neck—legally speaking—may be out a mile.

The real seriousness of the problem is present only on takeoff. Takeoff is when you are accelerating, when the engines are putting out full power, and therefore gasping for air (thus if something were going to "miss" the engine it might draw it in) and most of the incidents that I know of dealing with birds have had to do with takeoff.

It's a curious thing, but many people think that the landing is the crucial part of a flight. Actually, the opposite is true. Probably the most dangerous moment of any flight is just when you lift off the runway, because you are hanging on the engines. And if you lose power at that time you've really got your hands full. If you lose it on landing, you're not using much power to begin with—which leaves you with a lot more reserve on the remaining engine (or engines)—and on a good approach you are pretty well set up to land even if you lost them both in the last two or three hundred feet.

One thing that citizens should be aware of is that the site of an airport has a lot to do with whether or not birds, for example, will be a problem. If cities and states insist on building their runways near water and ponds, they can be sure of a bird problem. Logan Airport, well aware of the problem by now, has for years been systematically filling in the ponds near its main runways. Other airports should be doing the same. And those airports that are filling in the watery spots are actually adding an additional safety feature, for if a plane goes off a runway, it's a lot easier to get it and the people out of solid ground than a mud hole.

There simply isn't a whole lot more that can be said about the problem of birds. Some airplane manufacturers experimented with screens and other devices in front of the turbines, but they really didn't work. In some cases the birds (or other objects) simply drove the screens back into the engines. The airports do not, in general, do an effective job of keeping them away.

As for the birds themselves, one might wonder why they persist in returning, but as an airline captain said recently, "I guess they just figure that they were there first, and they have a right to stay there."

This is not, however, an area in which people should throw up their hands. In fact, according to one expert, bird control is one aspect of air safety where great strides could be made for relatively little money. For once, the technology exists to provide time protection, that is, such things as bird patrols that can get to the end of the runway and drive the birds away by using shotguns filled with noisemakers or live ammunition. Compare the cost of the DC-10 lost at Kennedy (where thankfully no people were hurt) with the cost of maintaining a small force of men, one or two vehicles, and a few shotguns. That's right. They don't really compare.

And it is also not as if no one is studying the problem. A Canadian group has been actively engaged in research for years, as have many other nations, including the U.S. Bureau of Fish and Wildlife. I find it interesting that London's Heathrow Airport has a staff of twenty-one, with four vehicles, that serve as a bird patrol. Apparently in this country neither airport management nor the government view this as the way to handle the problem. Perhaps it's too simple.

One other point to keep in mind is that birds will continue to be a problem until airport managers and certain FAA officials listen to the biologists (or hire some) who understand what it means to "sanitize an airport environment." What this means, in an oversimplified way, is that if you continue to build and run airports close to garbage dumps and beaches, you will continue to have potentially serious bird problems.

You certainly can't blame the poor birds. On a good day during migration an airport runway looks to a bird like a fine place to put down, rest, and eat. And in inclement weather, I'm told, a wet or icy runway apparently looks a hell of a lot like a smooth lake or pond.

As long as we keep destroying what the scientists call competing habitats, the birds will keep on coming. Must we wait for protection until the next time someone is killed?

* * *

Another important cause for concern that has been much in the news lately is the "near miss," the almost-crash of two airplanes hurtling through the air on what appears to be a collision course.

No other possible form of air crash so frightens people as the idea of a midair collision, and for just cause. It is rare indeed that even a single life escapes such a tragic occurrence.

This country's worst midair disaster occurred over the Grand Canyon on June 30, 1956. Two airliners—a TWA Constellation and a United DC-7—both flying at 21,000 feet, both legal, both in uncontrolled airspace flew into one another and all 128 people aboard the planes were killed.

[140] CLEARED FOR THE APPROACH

The CAB (Civil Aeronautics Board) investigated the accident and concluded, "the probable cause of this midair collision was that the pilots did not see each other in time to avoid the collision" and "there existed an insufficiency of en route air traffic advisory informations [sic] due to inadequacy of facilities and lack of personnel in air traffic control."

Fortunately, that accident remains the country's worst midair, or as one callous pilot described it, "the first of the famous midairs."

The airspace was getting more crowded, however, and there were several more midair accidents in the next few years. On December 6, 1960, two airliners crashed over Staten Island. (By great coincidence, once again it was a TWA Constellation and a United DC-7 plane.) This time both planes were flying on instruments, both under air traffic control, both receiving instructions from the New York Center. The TWA flight was cleared to descend to (and maintain) 9,000 feet, and when he reached that altitude the pilot was told to contact LaGuardia.

Once he did so the two planes were no longer under the same approach control, for the United plane was instructed to contact Idlewild Control. The two planes met at 5,000 feet. A total of 134 people were killed, 128 in the air and 6 on the ground.

As it turned out, following the investigation a dropline (a point-to-point, voice-activated "hot line") between the two control centers—something the controllers had been requesting for a long time—could have prevented the accident. Management had told the controllers working the scopes that it would be too expensive. The cost was seventy-five dollars a month. They got the drop line two days after the accident.

My experience with near-misses is not all academic, either. As I mentioned earlier, I once passed so close to an airliner that I could count its rivets.

It was back in November 1967, when I was doing a television show that took me into the homes of various people in the news. I had been up in Maine filming an interview with Jack Parr, and after the show was finished I gave a couple of the crew members a ride back to LaGuardia in my Cessna 310, and then headed over to Caldwell-Wright Airport in New Jersey.

FEAR OF FLYING

I had a defendant in a murder case waiting for me. I filed a flight plan out of LaGuardia, where there was a fairly low cloud cover, say tops of 4,000 to 5,000 feet, and was in contact with LaGuardia Departure Control getting vectors, as I was climbing on my way toward New Jersey.

Meanwhile, an Eastern DC-9 was coming into Newark on its way from Washington National. Actually, as I learned later, it was an empty plane being ferried up to Newark. The story, as it was related to me sometime afterward, was that the captain was preoccupied, and the copilot was handling the controls just as the plane headed into the soup. The copilot "transitioned his gaze from out the windshield to the instrument panel," and just as he—on Newark Approach Control—was about to enter the clouds, I popped out!

As I saw the thing go over me, I could literally count the rivets. We had to be less than 50 feet and probably closer to 20 feet apart. My immediate reaction was one of being distraught, swiftly followed by the state known as pissed off.

What had happened was that the controller handling him and the controller handling my flight were two different guys —and they were not in contact with one another.

Shortly after I landed in New Jersey, I was called to the phone and an FAA supervisor-type was beginning to read me out for being "a private pilot who got in the way of the big jet." As I had almost bought the farm through no fault of my own, I let the inspector know that I was not exactly a green pilot, and that in fact I would be glad to get the Eastern captain in there and find out what his side of the story was. They decided to drop the whole matter. But it was another example of a near tragedy that could have been averted by the presence of a seventy-five dollar a month telephone drop line.

Not long after these and other similar incidents, the FAA began to make plans for what is now the Common IFR Room in New York.

Near-misses have been an unfortunate fact of aviation with increasing frequency ever since the friendly skies began to get crowded. But no one ever has, or ever will have, an accurate statistic on just how often they occur because either one or both of the pilots—or the air traffic controller—may have been at fault; and, human nature being what it is, not too many report near accidents for which they might have been responsible, especially when the FAA might take punitive action.

One of the reasons why near-misses were in the news in the mid-70s was that the FAA had reinstituted its immunity program, whereby a pilot or a controller who reports a near-miss will not be "violated" or punished for his or her error. This program is used from time to time to try and get a picture of where the near-misses most frequently occur. Obviously, it cannot be permanent.

In late 1975 there were several near-misses of "high visibility."

One involved two jetliners, an American Airlines DC-10 and a TWA L-1011, carrying a combined passenger and crew load of 303 people. It was 7:30 P.M. on Thanksgiving, when a voice from the ground suddenly screamed at the American pilot to "dive." He did so with great suddenness and in 15 seconds his plane dropped 2,000 feet. Anything that wasn't strapped down, from meals to the people eating them, flew upward, and twenty-four persons were injured. But a horrendous accident was averted.

Less than two weeks later, on Friday, December 5, the same thing happened. This time it was two Boeing 727's, one owned by TWA and the other by United, and the crash was averted because the TWA pilot saw the other plane in time and took evasive action. As the wire service story said two days later: "The near miss Friday afternoon over Lake Michigan was the second time in nine days a jet aircraft came close to an aerial crash. A Federal Aviation Administration spokesman said it was the 14th near-miss this year."

Why? Well, it might be informative to preface the answer to that question by explaining that what the FAA spokesman meant was that it was the fourteenth *reported* near-miss of 1975.

I don't know the details of the second close call, but it was later reported that the first happened because the controller who had been handling the flights was finishing his shift, and he did not warn the man who took over about the converging patterns of the two aircraft. Fortunately, the fresh controller recognized the situation in time to scream a warning that saved over three hundred lives. It may sound horrible, but this type of near crash is not exactly rare. With the new equipment, the controller coming on duty does not have to spend as much time standing behind the man he is replacing to "memorize the picture."

A potential answer to this truly scary situation is the new device that is being introduced—albeit slowly—throughout the ATC system. It is a computerized warning that signals the controller that an airplane is leaving the airspace of one center and about to enter that of another. The warning consists of a flashing light that gives the number of the aircraft, thereby alerting the controller to pick it up and watch out for its safety.

Once the government decides to spend the money, all the busy areas will have this protection.

However, under the current system of traffic control, about one percent of the near-misses are unavoidable.

This is not to suggest that no one is worried about that small fraction. Both the FAA and PATCO have plans to remedy the situation. But the fact remains that it is still possible in this technological age for a midair collision to occur under circumstances where, legally, neither pilot is at fault.

* * *

There is another type of near-miss that deserves some discussion, but it differs in that it is a "planned near-miss." That is the lamentable situation (pointed out on the television news program *60 Minutes*) wherein a military plane uses a civilian aircraft as a practice target, by zooming up toward it and veering away at the last possible moment. In almost all of these cases, no one in the civilian plane—passengers or crew members—ever realizes that the military plane has even been in the near vicinity. But once in a while, as attested by the controllers from Jacksonville who were brave enough to tell the TV newsman all about it, the boys at the radarscope know what is going on.

The simple fact of the matter is that the military, either lawfully or unlawfully, will sometimes use a civilian airplane as a target for a "radar intercept." And if the pilot thinks he is a hot ticket he may even make a "gunnery run" at it. This is a practice maneuver used by peacetime pilots to simulate the wartime practice of trying to intercept an enemy plane while both aircraft are flying under instrument conditions, that is, both hidden by clouds or other weather conditions that keep them from seeing one another.

In almost all cases the fighter pilot will come up on the civilian aircraft from behind, pass under the belly, and zoom off. It's the kind of thing that could cause an observant passenger to age very quickly.

Just how often this happens is anyone's guess, for there is no accurate way of determining if it was purposeful or inadvertent, a clandestine mission or an airman's lark, or a frowned-upon-but-winked-at training maneuver. All I know (and Dave Savage tells me that his military experience bears out my recollection) is that it wasn't done when I was learning to fly jets. But then we had very little if any radar in those days!

I do remember that I was once called on the carpet for buzzing my mother's house. This was shortly after I had joined the reserve squadron. The Bedford, Massachusetts, tower called South Weymouth and reported that they had seen a Cougar—and we had the only Cougars around at that time—disappear below a 400-foot hill. The skipper called me in for a brief lecture during which he explained that somebody had buzzed Boston Harbor and the chief of Police of Quincy had jumped off his boat—therefore we were not to do any more buzzing. I got his message.

I don't mean to make light of the problem of military aircraft zeroing in on civilian flights. I feel that we should probably take the word of the controllers who first blew the whistle. And the FAA should not only listen to them, but make strong demands upon the military to cease and desist. And I understand that some letters of agreement are being worked out.

Finally, in the area of recent causes for concern, it is necessary to spend some thought in regard to the problems of madmen and terrorists.

Not that we can do a great deal about preventing the actions of a mad but lucid individual who is hell-bent on someone else's destruction. Witness the senseless bombing of the LaGuardia terminal in December 1975. But we can, as citizens, ask whether or not the government is doing everything it can to screen airports for the presence of the instruments of terror and death.

Everyone who flies should thank the natural or supernatural powers that be that the irrational pastime of skyjacking has apparently run its course. I don't think that we have seen the last of such incidents, but we are no longer at the mercy of any nut who thought he could get away with it—and be made a media hero to boot.

The establishment of airport luggage inspection facilities has certainly had a good deal to do with it, but we should remember that only carry-on luggage is subjected to electronic search. Perhaps we will come to the point some day where every piece of baggage that goes onto a plane will be examined. It would take a lot more time, but I'm sure that would be a better—and safer—idea in the long run. A little wait beats the hell out of an early demise.

Frankly, as far as the whole idea of terrorist-inspired acts is concerned, I think the less said the better. There is only so much that any rational being can do to offset the plans of a deeply disturbed mind. But the passenger should take comfort in the fact that the security system is here to stay. And if the need to increase that security becomes apparent, it will be done. So be it.

Another type of hazard that need not exist, and for which the government must be called to answer, was vividly demonstrated in the Spring of 1976.

On April 27th, at 3:10 in the afternoon, an American Airlines 727 carrying 88 people crashed while trying to land at the Charlotte-Amalie airport on St. Thomas in the U.S. Virgin Islands. Thirty-seven people lost their lives, and many others were severely burned. Although the recent National Transportation Safety Board report raises the possibility that the pilot was coming in too fast, the major contributing factor of the accident appears to have been the shortness of the runway.

The more candid airline pilots will readily admit that they do not like to land the big jets on strips shorter than 5,000 feet. The length of the runway at St. Thomas is 4,658 feet (plus a 500 foot concrete overrun) but that is not the entire problem: two good-sized hills adjacent to the strip create what is in effect a box canyon.

As one pilot who has flown the approach many times said, "St. Thomas is one of those airports where you have to put it right down on the numbers." (Meaning, you have to land almost at the beginning of the strip where the large numerals designating the number of the runway are actually painted on the asphalt.) Apparently, missed approaches are quite com-

[146] *CLEARED FOR THE APPROACH*

mon, for the decision to go in or go around—execute a missed approach—must be made at 500 feet.

On April 27th, the captain had all but landed the Boeing 727 when things began to go wrong. The power was off and the captain was about to "flare" the airplane, which means —without getting too technical—that he raises the nose slightly to facilitate the landing. Flaring, the last thing a pilot does before the wheels hit the ground, is a procedure far more easily demonstrated than described.

In any event, just as the captain was doing this he ran into turbulence. The force of the swirling air raised the right wing. A correction was made and the wings leveled again but the pilot simply couldn't bring the plane down properly. Had there been more runway, this would have been less of a problem, but the American flight was staring at a landing field that was disappearing fast.

(Later, the captain would tell the NTSB examining board that he felt he *did* have enough runway. And maybe he did have or would have if the air had been free of turbulence. But, like the pilot of TWA 514, he had more than one factor to contend with. His was not a position I would like to find myself in.)

He tried to take the plane back up in the air again for a very late go-around, but the plane was—according to observers in the tower—floating by then and he soon found himself in potentially fatal trouble.

The plane flew past the rest of the runway, past the concrete overrun, and struck a restraining fence that did not restrain it. It then rode up a ten foot embankment and snapped through the chain-link fence that marked the end of the airport.

After hitting a tree and a light pole, it smashed into the side of a filling station and raked to a halt against a palm tree in front of an office building. The inevitable fire eventually turned the aircraft into a total loss. For once the cockpit crew survived, but 37 others did not.

Within days of the crash, arguments sprang up about the length of the runway. FAA officials denied charges that the airport was unsafe, and a spokesman for American Airlines (which had just suffered its first fatality in 10 years) said, "We would not have been operating in there if we didn't think it was safe." He did, however, admit that the "safety tolerance is less than in other airports."

FEAR OF FLYING [147]

There was not really much else he could have said. If other airlines are going to continue to fly into that kind of an airport, "economic considerations" dictate compliance.

One particularly angry voice was that of J. J. O'Donnell, an Eastern pilot who heads the Air Lines Pilots Union. And it wasn't the first time he had bitched about St. Thomas. Back in 1970, following a similar crash at the same field, he had written the government and complained that the airport's runway was dangerously short. Two days after the 1976 crash, he told *The New York Times,* "There'll be another one there, no question about it. It's like playing Russian roulette with a gun with five bullets in it."

What made the crash so frightening was that the St. Thomas airport is one of three in the United States that received a "black star" rating from IFAPLA (International Federation of Air Lines Pilots Association). A black star rating means that instead of being either "deficient" or "seriously deficient," an airport's landing facilities are "critically deficient." Only two other airports are so rated—Los Angeles and Anchorage, Alaska.

As too often happens in air crashes, there were other factors that complicated the depth of the tragedy. On the day of the crash, the Port Authority at St. Thomas found itself, for some reason, inadequately prepared. There were two tanks filled with foam, but there wasn't enough water on hand to use in fighting the flames. They had to send for more!

The St. Thomas tragedy has at last moved the government to approve plans for extending the runway. Unfortunately, that decision is all too typical of this country's attitude toward air safety: hope that nothing serious happens, but if it does, *then* improve the faulty conditions.

No one concerned with aviation, from pilot to passenger, should have to stand for that.

N.B. The first of the court cases arising from the St. Thomas crash were tried in the spring of 1978, resulting in individual verdicts as high as 1.2 million dollars.

6

"After the Fall": The Trying of an Air Crash Case

Among the great pleasures of life for me are those that combine speed and beauty. But not everything that is both delightful and fast-moving is also fail-safe. Flying is a marvelous passion, but it has its Janus side.

Too often when a crash does occur, all one can do is pick up the pieces—literally—and from those pieces try to determine the probable cause, so that other airmen can be warned against duplicating that particular set of fatal conditions. All that is left then is to compensate the victims, wherever possible, at least insofar as the law is able.

When TWA 514 went down, killing 92 human beings, it had a direct effect on thousands of other lives. In some cases, the effect would be profound. For those who lost loved ones, their lives would never be the same again. And yet they would go through—because they had to—at least the first stages of a lawsuit bent on determining who was at fault, and who should receive what amount of money in exchange for a human life.

As might be expected, a field of study as complicated as aviation is not easily understood. Nor is it often mastered by persons who try to learn its basics in a short period of time. Add to that the standard complexities of the law, and you have a situation wherein expertise is uncommon. So it should surprise no one who thinks about it for a while that the "business" of aircraft litigation is one of the more sophisticated areas of the practice of law. As against the total legal population the number of aircraft specialists in the

United States is probably (in comparison with any other segment of the bar) minuscule.

The vast majority of air crash cases are handled by a small number of firms, most of them highly specialized, and most of whom employ at the very least one person who is truly an expert in the business of flying.

During the years that I represented the air traffic controllers, I considered myself to be disqualified from all legal cases involving airplane accidents, whether any ATC was involved or not, simply on the ground that they might be. When my tenure with that organization was finally concluded, I formed a firm in New York City, a limited partnership, to handle exclusively cases involving airplane accidents. Since the formation of that partnership, I have been involved in the majority of the air carrier disasters that have occurred during the last four years and a number of so-called general aviation accidents as well. (The one exclusion that I continue to observe is that of helicopter accidents, since I feel that as a manufacturer, it would be inappropriate and perhaps unfair to partake of the expertise that is available at manufacturers' meetings to turn upon my competitors in a court of law.)

Nonetheless, the insights afforded simply by being involved in air crash litigation are very similar to those that were once available when I was an investigating officer in the military, assigned to determine the cause of and the appropriate remedies for aircraft accidents that occurred during operations.

Infrequent as they are, the air disasters suffered by major airlines are the subject of massive litigation, whether they are midair collisions or the far more common collisions with the ground during a landing approach. And this extremely complex field is made more so, because in most cases conflicting state laws are involved. In the great majority of air carrier crashes, which are quite different in their handling from private air accidents, the injured parties (and in cases of death the survivors who have a legal interest or responsibility in the estate) will consult with the family attorney as to what if anything should be done to remedy the damages that have been suffered. A large percentage of the practitioners who are greeted with such problems will immediately recognize that the business of airplane accidents is far beyond their legal expertise, and by professional reference will locate a law firm with experience in such matters to take over the responsibility for whatever litigation they think appropriate. In some cases, unfortunately, a local practitioner will attempt to handle the matter himself, and though he may do a creditable job with no prior experience, in most cases he is doing his client a disservice.

Those who are in the aircraft accident business on a regular basis shake their heads in dismay whenever a local lawyer who wants to avoid a referral fee will (with almost no investigation, preparation, or expenditure) accept a settlement far below that which other passengers are getting, just to dispose of the case; or, worse, where families who have not even consulted counsel and are particularly ignorant of whatever rights they might have are visited by the friendly insurance adjuster for the air carrier responsible for the crash and induced to accept rather picayune settlements which are described to them as extremely generous—and which any experienced lawyer would recognize as grand larceny.

The actual litigation of an air crash case is a long and drawn-out process. First, it is necessary to identify not only who *is* responsible for the crash, but also who *might be* responsible as well, and herein lies a tricky phase of the entire business.

An obvious and immediate target of any lawsuit claiming damages for death or injury resulting from an unsuccessful flight is the air carrier itself; however, the search does not end at this initial point. If, statistically, most accidents are the result of pilot error and therefore the responsibility of the air carrier that employed (and thereby vouched for the pilot), there are a fair number that may arise from other causes.

Potential defendants in any crash suit include the manufacturer; the overhaul facility if one was used to overhaul major components like engines, props, and so forth; and the FAA itself, if a controller has made a mistake that caused the accident to occur, and particularly if a midair collision is involved when both planes were under positive control. The government could also be liable for the failure to properly maintain the radio aids upon which a pilot depends for the safe conduct of his flight, including en route navigational aids, but more particularly those that are used for his final approach where for instance there is a known malfunction in the instrument landing system (ILS) or the VOR or ADF from which the approach is shot and no appropriate warning is given to those who use these facilities.

There has also been suggested, although it is a vague area of liability, that when the government grants to a manufacturer a certificate to produce and sell a certain kind of aircraft—while overlooking a patent defect in the aircraft which ultimately causes some kind of air disaster—that the government together with the manufacturer should be liable for the consequences.

In the main, these are the principal targets of any aircraft lawsuit, although in bizarre circumstances others who have something to do with the operation of the aircraft could conceivably be brought in.

One of the principal complexities in the lawsuit (or lawsuits) that results from any crash of a public air carrier derives from the different citizenships involved, that is, from state to state. (When citizenship of foreign countries is involved, the problem is even more complicated.) Although there has been a general leveling of legal practice in many sectors of the law in an effort to minimize the impact of state lines of the rights of interstate travelers and businessmen, one of the principal "survivors" of this general amalgamation of rights has been the measure of damages in the event of death.

Here the variance travels clear across the spectrum. In some states of the United States, a human life is worth a few thousand dollars. In other states, it can be worth millions, depending on the identity and the position of the decedent. Naturally, every litigant wishes to take advantage of the law which most favors him or his decedent. And toward that end, a great deal of so-called forum-shopping is done. The legal concept of forum-shopping is no more than seeking that court which will give the greatest advantage to either party to the lawsuit.

One of the most frequent users of the federal system of "multi-district litigation" is that segment of the bar which specializes in the handling of aircraft cases. Ordinarily, in the wake of an air disaster multiple suits will be brought in different jurisdictions, each by plaintiffs wanting to vindicate their rights in the most favorable legal atmosphere. Since the law attempts to discourage trying the same basic case ninety times—if that's the number of passengers who were involved—there is a provision in federal courts, where most of these cases are litigated, to consolidate all of the lawsuits before a single court which will have the power to determine who is liable and the further power to determine what each plaintiff may receive as his damages in the case. This, as to individual plaintiffs, can vary quite widely.

A federal court sitting in New York, following a New York crash of a plane which had departed from Dallas, for instance, could well decide that some of the occupants were entitled to whatever measure of damages Texas law would allow, some would be permitted whatever New York law would allow, and some who had boarded the flight in Los Angeles and only connected through Dallas might be permitted whatever California law would allow.

Because of the vagaries of what lawyers refer to as conflict of laws, which is simply the different kinds of laws from different states being in competition for application to a particular lawsuit, the expertise of the firm in charge of the case becomes rather critical. A mistake in seeking the proper remedy could result in a loss to the client of hundreds of thousands of dollars, simply by making him subject to a limitation which could have been avoided if proper legal steps had been taken.

All of this assumes, of course, that there is no difficulty in establishing that *someone* who is financially responsible can be pinned with the blame for the accident. In many cases this is not very difficult to do. In other cases, where it is perfectly apparent that someone has to be responsible, it becomes rather difficult to establish which of several potential defendants ought to shoulder the principal part of the blame. Here again, the expertise of the lawyer involved is called into sharp focus in order that a defendant who should ultimately be held, not be released prematurely on some small settlement, only to discover at a later time that the other defendants are being excused one by one on the ground that the party who should have picked up the lion's share of the tab has been excused for a menial amount.

In some cases it is extremely difficult to show that there is any liability at all. These cases generally revolve around a dissonance between defendants, each of whom points the finger at the other (in a perfectly plausible way) and assigns to the other the total responsibility for the accident. And there is an even worse situation: those cases where the cause of the accident is attributed primarily to that target which has never been held accountable to the courts of the United States, called in the law "an act of God." Such acts may encompass sudden and unforeseen weather conditions which no pilot could reasonably have anticipated, air traffic control and the National Weather Service could not reasonably have warned of, and which are admittedly so destructive of flight conditions as to make their confrontation beyond the pale of handling by even the most expert airman. When these conditions prevail, they strongly suggest that the plaintiff—no matter how badly he may have been damaged—is without remedy.

The investigations that are involved in air crash litigation are among the most extensive of any lawsuit ever pursued in the United States. It is not uncommon for the depositions (that is, the question and answer statements taken under oath from the many witnesses, both inside and outside the aviation field who have any knowledge, even peripheral, of how the crash occurred or what was going on as it was in progress) to stretch for 40 or 50 feet. Needless to say, the expense involved in assembling this material is very, very substantial; and it can be borne only because most of the firms that go into this sort of thing have several clients to represent and can anticipate a fair degree of success some years down the road. Thus, they can foot the bill for the costs that are necessarily incidental to the proper preparation of a difficult air crash lawsuit.

There have been instances in the past where determined lawyers have actually gotten themselves hired by the manufacturer suspected of being responsible for the crash and worked in his plant long enough to find out exactly what the (hotly denied) defect was that caused the crash in the first place.

One well-known air crash lawyer went to work for six months in the plant of a large parts manufacturer. He not only learned exactly why the part had failed, but when the case went to trial, he (to borrow a seldom-used phrase from the lawyers' grab bag of jargon) "hung them from a tall tree."

Statistically speaking, there are many more small airplane crashes ("nonair carrier") each year than there are among the airlines. Although the number of victims in each instance are nowhere near as large, in all probability more court hours are consumed in litigation of the so-called small crashes than those that involved certificated air carriers. At the same time, the burden upon the plantiff's counsel becomes somewhat greater. Whereas the air carriers owe a very high degree of care to their passengers (as do air taxi operators who fly passengers for hire) the fellow who goes out with a friend on a Wednesday afternoon to fly from Cleveland to Cincinnati has a heavier burden to shoulder in order to collect his damages. He also has more potential defendants to consider: in addition to the manufacturer of the aircraft, the company that does the maintenance, and the pilot responsible for the accident (if he was), it is also necessary for him to make several other inquiries. He must attempt to learn whether the fueler who last filled the tanks put the right grade of fuel in and whether or not it was contaminated with water; whether the examiner who last gave the pilot his currency check in fact gave him a proper and thorough check and was qualified to give it; whether the line boy who preflighted the aircraft took all of the appropriate steps before declaring it safe for flight, and whether the hangar-keeper who housed the aircraft permitted it to suffer some sort of "hangar rash" or minor collision that might have contributed to the accident.

The possibilities of responsibility are almost endless in private aircraft. And although the complex legal problems which almost invariably arise in air carrier cases may be less onerous, the imagination of counsel in putting together a case is frequently drawn upon somewhat heavily.

There are several things going for a thorough investigation, not the least of which is the immediate inquiry conducted by the National Transportation Safety Board in every serious accident, be it commercial, private, or other. All of the physical evidence, which is frequently strewn across several acres of ground, is immediately the property of the NTSB, and without the normal legal impediments to a thorough inquiry, its officers are allowed to move forward and investigate every possible aspect of the crash. Although an alert lawyer who is retained in sufficient time to become involved in a threshold investigation of an air crash has every possibility of being invited to attend the inquiry, in many cases the NTSB has concluded its investigation before counsel has even the first crack at learning what went on.

Most lawyers in the aviation field have a very high regard

for the efficiency and the expertise of the NTSB; on the other hand, very few lawyers who have experience in the field are willing to simply close the file with the National Transportation Safety Board's finding and recommendations. There are several reasons for this. One, it is not the job of the NTSB to decide who is responsible and therefore should pay whatever damages may have resulted from an air crash; its official responsibility is to determine, as best it can, the causes of a crash and to make recommendations which will alert other pilots to avoid those same causes in the future. Secondly, the Board's investigators are not lawyers trained in the business of attempting to make recovery for some bereaved widow who is left behind with five children, four dogs, and a heavy mortgage. Furthermore, it is not the board's responsibility to line up and identify all peripheral or adjacent defendants who may have had some participation in the responsibility for the accident.

In the main, the NTSB tends to fix responsibility on the pilot and to designate the cause of most accidents as "pilot error." Even though this finding may be principally correct, the rule of law requires that all parties who contribute to an accident share in the responsibility for whatever harm has been caused, so saying that the pilot made a mistake is no end to the matter. It is the responsibility of the lawyer who is hired to recover whatever damages the law may permit to not only find those who were responsible for the accident, but also those who were responsible *and* have the ability to make some form of restitution.

In all too many cases, the pilot who is assigned the primary responsibility has left—if any—a rather modest estate (pilots are not usually rich men) so the lawyer must find those contributors toward the pilot's woes that can be held to account for what has occurred. In essence, this puts upon counsel two rather distinct burdens: one, he must enjoy some degree of expertise in the field of aviation; and two, he should be a pilot of some experience.

To the plaintiff who is bewildered by the entire situation and wonders with some consternation, "Who if anyone is going to pay the cost of the loss?" a pilot-lawyer is a pretty good bet. The reason for this is that pilots generally react against an assignment of "pilot error" as the prime or only cause of the accident. Although there are some cases in which no other cause can even be suggested, in most, the pilot who meets an unfortunate end has encountered some obstacle that he didn't expect, and which, at least to him, contributed very seriously to his failure. Pilot-lawyers tend to take this view and to search diligently for those other factors which they—if they had been in the same confrontation as the decedent pilot—might have viewed with a good deal of trepidation and resentment.

The history of air crash litigation, particularly in the private sector, has involved tireless attorneys with some considerable degree of expertise in the business of aviation who have been willing to probe, look, test, recreate (in some cases all but *duplicate*) the hazard involved to attempt to find out who and what was responsible for the accident.

The next step is to parcel out the blame and see who should pay the damages. In many instances, this necessitates a trial.

In early December 1975, just before the trial in the cases arising out of the crash of Flight 514 were about to begin, the FAA and TWA announced that they had reached an agreement: they would share the liability. Each party would admit to 50 percent of the blame. By so doing, the two parties avoided a long and costly court battle over this central aspect, which meant that the only issue to try would be that of damages.

If this sounds somewhat strange to you, that is because the trial of an aircraft case is different from the vast majority of civil or criminal trials.

If the government is the sole defendant—say the case involved alleged controller error or faulty equipment at a federally owned airport—then the case is heard by a United States District Court judge sitting alone and without a jury. Neither side has a right to a jury trial. (When Congress passed the Federal Torts Claims Act in 1946, it felt in its "wisdom" that the juries hearing cases against the government might tend to favor the plaintiffs. As one attorney with a great deal of experience in federal court in cases against the government put it, "Congress was afraid that juries would give the farm away.")

In cases where there is more than one defendant, such as the TWA 514 cases, the situation changes and there may be what is known as an advisory jury on the question of the government's liability.

A government lawyer once told me, "You're going to have a very poor track record from a won-lost standpoint on aviation litigation if you're representing the defendants, because in addition to the typical sympathetic attitude that you always have in a case like this, the compounding effect of multiple deaths will lead a court or a jury to almost any lengths to find some theory of recovery upon which they can award money."

Another reason why air crash cases usually are (and always should be) handled by specialists is the complexity of the legal theories that develop. All aircraft cases come under the heading of "torts," which is the law's generic term for distinguishing between criminal acts and harm done by accident. A tort is a civil wrong. It can be the result of a conscious act performed in a careless manner (like tossing a lit cigarette out of a car window, which starts a fire in a passing convertible) but without any intent to harm; or a tort can be the result of an omission, a failure to act (like digging a deep hole in your lawn and then forgetting to cover it at night). But it is not a criminal act, nor does it cover legal actions that stem from breach of contract.

Law students usually love their course in torts, because it is one of the few areas of the law where humor breaks through with some frequency. I would guess that 90 percent of the lawyers practicing today remember the classic definition of "negligence per se"—the textbook case wherein a person sued because he had bought a tin of pipe tobacco, opened it, and discovered a human toe. In its opinion, the court wrote, "The presence of a human toe in tobacco is negligence per se." I have my own favorites, though. One has to do with the man who brought suit because during the heated progress of a wrestling match he was injured when his opponent "too vigorously" squeezed his private parts; the court ruled against him, finding that by taking part in the wrestling match, he gave his implied consent to the risk. There was also the case of the woman who went to the circus and loved it—until a horse backed up to her front-row seat, and "evacuated its bowels in her lap." She sued for mental anguish as a result of the accidental injury and she won.

Air crash cases however are seldom humorous, and they are rarely simple. Imagine the work involved in trying to show that the crash of a small plane was caused by the wake turbulence of a large jet. Witnesses must be put on the stand to testify regarding such esoteric knowledge as "wing tip vortices," the record must be searched to indicate the exact moment the controller allowed the smaller craft to begin its final approach, and any number of other possibilities must be considered and then used or discarded. Yet that is by no means the most difficult example I could have chosen.

In fact, about a dozen years ago the Justice Department admitted that the lawyers in its torts section were having a great deal of trouble defending the government in air crash cases. There were cases where the government was not at fault, but it was held liable because an able, experienced plaintiff's counsel knew what he was doing and the government lawyer didn't.

To remedy that situation, Justice set up a separate division and recruited lawyers to become specialists in air crash litigation. It had to, for the sudden rash of major air accidents in the early 1960s had brought a lot of suits into court. (One might remember from the previous chapter that this was a few years before the controllers all but threw up their hands and formed a professional association. Air travel might have been booming, but air safety was not.)

Today, the men and women who represent the government in air crash litigation do not have to turn around and try a land condemnation case the next day. And that's the way it should be.

Another problem caused by the complexity of air crash cases is that it takes one hell of a bright judge to cut through the thicket and follow the case. This is not a unique problem, however, for few judges are equipped to follow the nuances of a major patent case, for example, but at least in that area of the law there is a special court of appeals one step below the Supreme Court. A few years back a lawyer told me he had been involved in a large air crash case, and the evidence and procedure got so complicated that the judge—an unusually bright judge—began each morning's session of the trial by shaking his head and saying, "Gentlemen, I'm never going to fly again. I'm never going to fly again."

In October 1975 I gave a speech in Milwaukee, Wisconsin, and I later met several lawyers over a drink. One of them, Michael Wherry, turned out to be one of the first lawyers hired by the government—right out of law school in the early sixties—to specialize in air crash litigation. We had an interesting time comparing notes, but he made one point that I could not have agreed with more. That was the effect of the complicated and complex trial procedure on the air traffic controller who may or may not have been at fault. It pleased me to see that a government lawyer, a former government lawyer, had seen way back then that the controllers were under an unnatural amount of pressure.

As Wherry put it, "I never did feel that the individual controller was the snag in the system, not by any stretch of the imagination.

"In contrast to someone who drives a bus, for example, who is seldom going to find himself involved in a major accident case, the controller is subjected to the massive publicity, the amount of time that's involved in preparing a case, and all of that has to give most people a feeling of second-guessing, perhaps even guilt feelings, whether they were warranted or not.

"I could see this in the faces of these people, working with them, having a drink with them afterwards, trying to get them squared away on how to cope with a very skilled lawyer, that their confidence was shaken. And the sad thing is that the confidence is what makes them able to do this extremely unusual, high-pressure job."

* * *

The professionalism of the lawyers who try air crash cases, and I'm talking about both sides, is much higher than one finds in either civil or criminal litigation generally. Certainly one of the reasons for this is that the difficulty of the work attracts able counsel, but there is also the undeniable but omnipresent fact that we are all caught up in the aftermath of a tragedy.

But too often we come dangerously close to injustice because of the simple, nonjudgmental factor of red tape. Airplanes, despite this country's marvelous record for air safety, do crash. And when it happens to you or your friend or family member, you cannot be consoled by statistics. You can only be paid a sum of money that a court decides will be fair, given the circumstances.

CLEARED FOR THE APPROACH

The horrible irony is that you may receive a lot more—or a lot less—because your husband or wife or mother or father happened to decide to live in one state and not another.

There is nothing just about that.

PART II
LOVE OF FLYING

1

Love of Flying

Nearly every month of my life, for almost as many hours as an airline captain flies to earn his salary, I can be found in either the pilot's or copilot's seat of something that flies. In many senses of the word, I fly for a living—but it is my own living for which the flights are made, not someone else's.

Nearly every interviewer I have spoken with during the last ten years has put the same question: "How do you do so many things at the same time and keep up with it all?"

My answer is a simple one: "Excellent staff and associates *and* fast airplanes."

The whole truth, however, goes deeper than that. Fast airplanes in and of themselves would never be enough if the hours aloft were agony, anxious, or even boring. It is the ability to relax totally in an atmosphere of sheer delight which makes an interstate and occasionally international life-style tolerable. And I suppose the key to enjoying this state of mind while whisking from here to there is the peace I feel from having a complete understanding of what is going on, the sensation of complete control of my environment.

Mine are not merely fast airplanes, but superbly equipped as well. Indeed, in the view of many, a glance at any of my instrument panels might cause the reaction that fitted there is everything but the kitchen sink, and two of them at that. Damned right. Man has never devised any instrument or machine that can't break down just when you need it most. I even have two of the things *most* likely to be the primary factor in any accident—the pilots.

But there is a very good reason for all this redundancy of men and machinery. My life is a series of commitments, very firm commitments, whether they be court appearances, speaking engagements or business meetings. In all but a few of these a no-show is simply unthinkable, which means that when a trip is scheduled, it goes. It goes no matter what, and the "what" may be rain, snow, ice, fog, turbulence, thunderstorms, or some combination of these. It goes when the visibility is so low that the air carriers have been grounded, and it goes to places they have never been. It goes on schedule and arrives on schedule, safely, every time. Looking back over the past ten years, and the thousands of flights I have made during that time, I cannot recall enough "aborts"—instances where we did not arrive as planned—to count on the fingers of one hand.

This is not the product of extraordinary good luck, or the result of a direct line to the Almighty's weatherman. Indeed, there are many business fliers who can point to similar records. It is simply very good evidence that given a well-designed, well-maintained, well-equipped airplane, and a pair of confident, seasoned pilots who have carefully planned every phase of the flight, air travel through our system is not only safe but very, very dependable.

When I was in law school and spending most of my time doing investigations for trial lawyers, I was given an assignment to take some aerial photographs of some land that was in litigation. I asked the lawyer scheduled to try the case if he would like to come along, to get a firsthand look at the subject matter of his lawsuit. He quickly declined saying, "No, thanks. I'm a great believer in terra firma, and the more 'firma' the less 'terra.'"

And yet my barrister friend was hardly a conservative soul.

He had no compunction about driving at eighty or better along the interstate in heavy traffic, where a blowout or a drunk could spell instant disaster which he could neither avert nor control. I never saw him punch the starter of his twin-screw motor yacht without first clearing the engine compartment of fuel vapors, which, if not done, is a good way to set off a major explosion. He wasn't a fearful or an anxious man. But he was terrified of airplanes.

And there are still, in an age when flying is more of a routine business or vacation necessity and no longer a great

LOVE OF FLYING

adventure, many such people. A few, perhaps, have genuine phobias arising from deep-seated psychological problems. Most that I have known suffer from an attitude that is as old as mankind itself—they have a dark and innate distrust of anything they cannot fully understand. They will listen at length to persuasion that wings do generate lift and that engines are designed to keep running, but they are hesitant to believe it. I think that if I felt even a flicker of this sort of doubt, traveling would become an ongoing nightmare, and I might well become a very local person.

I am sure that I enjoy and appreciate the world of flight more than many because I am the victim of an intense curiosity, and must seek out a thorough understanding of anything in which I am involved. And there is no better place for a curious young airman to find himself than in military flight training, where a thorough understanding is not only available, but necessary for survival. Simply put, if you're going to try to plunk a high-performance fighter plane on the pitching deck of an aircraft carrier after flying a blind approach through the soup, you had damned well better know what you are doing.

Having been blessed with an excellent education as a pilot —and for my money there is nothing to compare with the flight training offered by our armed forces—I have managed to poke my nose, for one reason or another, into almost every phase of aviation. Aircraft accident investigation and litigation, grisly business though it is, provides a solid opportunity to avoid a bucketful of mistakes that other pilots have demonstrated to be serious.

Standing in the control towers and radar rooms with the air traffic controllers at peak work load and attempting to help them work out their problems during a very difficult period in their development and growth has made working with them from the cockpit infinitely more easy and pleasant. Operating a flight school, charter service, and airport—and trying to make ends meet in the process without stretching safety—affords insights into the business of flying that many pilots never really see. Several hundred hours on the flight decks of major trunk air carriers talking with the crews and watching them work affords a direct view of millions of flight hours of experience, from which a great deal can be learned on a daily basis. And finally, involvement in the design and manufacture of the machines themselves, with all of the excitement and responsibilities that are always present, affords an understanding of the very roots of the business.

With this kind of rather broad perspective of the world of flight—and I consider myself extremely lucky to have been able to acquire it—it is very easy to feel secure aloft, confident that any malfunction in any part of the system can be noticed, understood, and compensated for quickly and safely. In that frame of mind, I find it quite comfortable to put the worries on the back shelf and devote my attention to all that is pleasant about being in an aircraft.

And there is a good deal that is very pleasant indeed. From the moment the checklist has been completed and the throttles are pushed forward, there is for me an exhilaration in the whole experience unlike anything else in life. Feeling that "boot in the ass" as the engines come up to full power, watching the airspeed indicator wind up toward liftoff speed, and easing back on the control column as flying speed is reached, hurtling past the barrier of gravity and into a three-dimensional world somehow promotes the notion once again that life isn't all that bad after all. Problems seem to diminish somewhat in proportion to the shrinking of the landscape below.

To wheel merrily around and over the puffs of a clean white floating cumulus cloud, or to slide effortlessly through it, is to be reminded that this machine is yours to control, to do your bidding, to take on whatever the weatherman can conjure up without hesitation or complaint. If dreams are to be dreamed, they can best be envisioned by looking to the horizon where infinity is found and the realities of life on the surface are no constraint. A private airplane is, in its best sense, a very private world.

When we are leveled off at cruising altitude and the air is clear and smooth, I often leave the management of the machine to my co-captain and "George," the autopilot. In this situation the work load is very light, with only an occasional button to push, radio frequency to change, or controller to check in with. The passenger cabin, suspended between the comforting sound of the engines, is a great place to unbutton your collar, stretch out, and relax and let the hurried pace of life wind down to an idle.

If there is someone seated with you, it is because you wanted that person there, not because some stranger bought a

LOVE OF FLYING [167]

ticket. There is an air-to-ground telephone, but it is equipped with a marvelously simple little switch which moves to Off. (It would hardly do to have the pilot screen incoming phone calls by saying, "I'm sorry, Mr. Bailey just stepped out for a minute.")

If there are reports or files to be studied, there is no interference or distraction. If a good novel is at hand, total immersion is easy to accomplish. And if the day has been or promises to be a hard one, there is a great hour or two or three of sleep to be had.

If the destination is CAVU (ceiling and visibility unlimited), George and the man up front don't need any help, and it is pleasant to sit back and simply observe the descent, approach, and touchdown. If there is bad weather or some other problem, then it is time for teamwork. And teamwork in an aircraft cockpit, smoothly coordinated and orchestrated, is probably more precise than that of any sport or profession.

A packed instrument panel—one on which every square inch seems to be occupied by a switch or a gauge—will befuddle the mind of the average layman. He is apt to say, "I could never keep track of all those gadgets!" But to a pilot the array of information presented by the multitude of gadgets makes the difference between marginal safety and good solid control of the situation.

When I learned to fly in the early fifties, the sophistication of our present navigational radio system was just being discovered. We listened to all sorts of weird sounds emanating from the old low-frequency radio range, tried to guess where we might be, and attempted by constantly scanning different instruments to keep the aircraft reasonably upright.

But the present state of the art has greatly simplified all of that. One glance at the panel is enough to indicate one's exact position over the ground, and a single instrument—an electronic wizard called a Flight Director—gives enough information to control the airplane. For full redundancy, both the pilot and copilot have a complete set of everything. Should a critical instrument fail during an instrument approach, the other pilot simply takes over.

Although flying "on the gauges"—that is, in clouds, fog, or other zero-visibility conditions where control is possible only by reference to instruments—is anything but fun in a plane not fully equipped for the exercise, for me at least the reverse is true when the goodies are before you. When the cocaptain turns around and points to his kneeboard, I know that he has just received the current destination weather, and that it is what the Sunday pilot would call "not good." But so long as the field at which we are landing has minimums —usually a ceiling of two hundred feet and visibility of a quarter mile—the signal simply means that the fun is about to begin.

And flying instrument approaches *is* fun, so long as the crew stays with air traffic control and ahead of the airplane. It is an exercise, a drill, a chance to demonstrate to yourself once again that so-called bad weather is a "piece of cake." And the fun is there whether you are in the left seat actually controlling the aircraft, or in the right seat handling the radio and the rest of the copilot's chores.

The first step in preparing for the approach is to set the geography of the destination firmly in the memory by studying the Jeppeson Approach Plate very carefully. This in essence is a small map, printed especially for a particular kind of approach to a particular runway at a given airport. It depicts the field itself, the radio beams which are to be used, the frequencies of all of the different ground aids that must be tuned in, the field elevation above sea level, all obstacles (particularly radio and TV towers) showing their height above the ground, and the minimum altitude that must be kept at various points in the approach. It will also indicate if there are any special features available to make the job easier— items such as distance-measuring equipment connected to the instrument landing system, precision approach radar enabling the controllers to "see" the airplane throughout its final descent, and high intensity runway lights.

These lights are frequently the pilot's best friend. They are a series of electronic strobe lights, similar to the flashguns used by photographers, which are timed to flash in rapid sequence. As seen from the cockpit, often through dense fog, the effect is one of a very bright light streaking toward the touchdown point. For some reason unknown to me, the high intensity light system has been dubbed the Rabbit, and every pilot who is nearing the minimum allowable altitude and reaching for the throttles to "go around" has warmed to the sound of his copilot saying, "I have the Rabbit running." Unless the Rabbit disappears again, it is usually safe to land.

Once that appropriate approach plate has been memorized, the checklists are run. In well-disciplined cockpits the copilot will read the checklist aloud and touch each switch or control with his hand as he announces that that item is "check." Some pilots disdain checklists, preferring to use the memory instead, and many of these get away with doing so for years. After landing one time with the wheels up, they will start using the checklist again.

As the plane gets close to the outer marker—a point about five miles from touchdown where the final approach is begun—the radar approach controller in the radar room below the control tower will be heard to say, "Cleared for the approach." For perhaps the first time in the flight, the pilot is on his own. These are the words that complete the actual "control" by the men on the ground. No more will they give instructions as to what direction, speed, and altitude to fly. Unfortunately for too many people "cleared for the approach" is the last phrase they ever hear.

Once that final clearance is obtained, it is time for the pilots to "come up to speed"—to hone their senses very carefully, to scan the instruments with disciplined regularity, and to talk to each other, checking and double-checking to see that all the instrument indications are showing what they should show and are in harmony with each other. It would sound strange to the uninitiated, but in essence the pilot who is actually flying the approach is being told by his copilot what he is seeing with his own eyes. Or at least he had better be hearing what he is seeing, because the copilot is describing what appears on an identical but completely separate set of instruments, which should give readings exactly the same as those which appear before the pilot. Should the pilot pick up any discrepancy between what he is looking at and what he is being told, he is going to have to find out which side has the malfunction in one helluva hurry.

The last phase of an instrument approach, and the one where mistakes can no longer be tolerated, begins when the outer marker is passed, an event which would be difficult not to notice even if one were sound asleep. All kinds of things happen at once. At a point fifteen hundred to two thousand feet above the ground, about five miles from the end of the runway, a vertical radio beam from the ground causes a blue light to flash, a buzzer to buzz, and the pointer of the automatic direction finder radio to swing from the nose to the tail. At the same time the glide slope—a slanting radio beam angling down at about three degrees and leading directly to the runway touchdown point—is intercepted. The wheels are lowered. Approach flaps are lowered. The throttles are retarded to reduce power for a descent of about five hundred feet a minute. And the tower is told, "Outer marker inbound." The checklist is run once more.

If the airplane is a jet, or a large business twin, there is just a little more than two minutes left to fly, and in that two minutes there is no room for sloppy air work. Unless the pilot keeps his localizer needle—which reads a radio signal directly aligned with the runway—right in the center of his instrument, he is going to miss the runway on one side or the other. If he allows the airplane to drift below the glide slope and fails to correct smartly, he is going to hit something other than a runway and make a mess of things. If he flies above the glide slope, he will break out of the clouds beyond the touchdown point, and have no room to land.

The copilot is also a busy man. In addition to monitoring his instruments and reporting their readouts to the pilot, it is his responsibility to keep glancing through the windshield in search of terra firma. Unless he says, "I have the rabbit running," or "Ground in sight, go contact" before the plane is at two hundred feet above the surface, the pilot is going to have to (very smoothly and suddenly) add power, lift the wheels and flaps, raise the nose to climb, and tell the tower that the approach has been missed so that he can be switched back to radar control. The copilot must also keep a sharp eye on what the pilot is doing and warn of any mistakes in progress in clipped, even harsh, language. The pilot may have the controls, but crashing is togetherness.

As I have said, thanks in great measure to the fact that I have found very skillful people to fly with, a missed approach has been an extremely rare event. And making it through rough, greasy weather is its own kind of thrill and satisfaction. If we land when some of the air carrier pilots are taking missed approaches and going to alternate airports, the satisfaction is boosted just a little, for the airline pilots are the measuring rod for flying skill.

And when conditions are tight, there is usually someone in the terminal who has been chewing his or her nails, worrying that our airplane won't make it in. It may be a client who is due in court, a convention program chairman who has a crowd waiting for a speech, or a businessman who has a deal that *must* close by four o'clock. It makes little difference. They all beam the same smile of relief.

And the feckless pilots, of which I am always pleased to be one, who minutes before were sweating bullets over whether the Rabbit would appear in time, are very quick to shrug off the fact that they have defeated the murk which enshrouds the airport with, "routine landing, strictly routine, all in a day's work...."

* * *

My own experience with flying dates back to 1952, when I entered Naval Flight School at Pensacola, Florida, during the Korean conflict. For eighteen busy months I endured and enjoyed a superb and rigorous training syllabus—to my mind the best ever devised.

Almost immediately after receiving my wings, I hustled down to the local office of the Civil Aeronautics Administration (now the FAA) to take a written examination for a commercial pilot's license.

Just a few days later, I had an interesting and somewhat mind-boggling experience. Having flown nothing but heavy high-powered military trainers and World War II fighters, I took my first lesson in a jet one morning and my first lesson in a light civilian trainer that afternoon. The jet went okay, but the little plane was a handful; on my first approach, I floated over the entire length of a mile-long runway without ever touching down.

A month later, as a jet-qualified Marine Second Lieutenant, I was stationed at Cherry Point, North Carolina, with a fighter squadron. I immediately bought a little Luscombe two-seater for one thousand dollars and tore it all apart. After an engine overhaul, and new paint and interior, I put the Luscombe into the air during every off-duty minute I could find. Although not specifically directed by the North Carolina authorities to do so, I felt it my duty to make frequent patrols along the beautiful desolate Outer Banks to make careful tally of those sun-worshipers who had forgotten their bathing suits. On rainy days the veteran bush pilots of the Beaufort-Morehead City Airport would enthrall me with one exciting tale after another (slightly embellished, of course) about flying. To one whose exposure to hairy tales had been strictly military, this was a whole new world.

In the steamy summers of the middle fifties, I learned a lot about the personal affection a pilot develops for his airplane. Cape Lookout, a few miles from the airport, became a favorite target for huge hurricanes which spawned in the Caribbean and thundered ashore on the Carolina coast. Unable to afford a hangar, I would dig holes in the ground for the landing gear and chain the Luscombe down to large stakes. During the twelve hours or so that each howler would batter the little airplane with winds as high as one hundred fifty miles per hour, I would sweat like some parent whose child was out in the maelstrom. In one particularly violent storm the Luscombe suffered minor damage and every other plane on the field was all but destroyed.

In the summer of 1955, I was the legal officer for Marine Aircraft Group 32, the parent of my former squadron. We went through a change of command as Col. Robert H. "Blackjack" Richard was relieved by Col. Stanley W. Trachta, and I was delighted to serve on staff to two first-class marines in succession. One day Col. Trachta called me in. "Lee," he said, "I've been thinking about buying an airplane—you know, something to fly on weekends. But I figure that what I really need is about *half* an airplane, with the amount of time I'll have to use it. At the little grass strip just outside the base there happens to be a real clean little Swift which is for sale at a right price. Would you be interested?"

I gulped a little. I knew the airplane well. It had belonged to my friend Charlie Vellines, an overhaul specialist and CAA inspector, who had recently sold it to buy a four-place Stinson that he could use to go fishing on the Outer Banks. It was immaculate. But a mere lieutenant with one and one-half airplanes? Oh well, I could always sell the Luscombe. The deal was on.

The Swift, although not really very fast, looked like a P-51 fighter. We flew it religiously and ultimately gave it a complete face-lift and a new and larger engine. We had paid seventeen hundred dollars for it, a bargain induced in part by the fact that a hurricane was headed for the area and the seller had no hangar. When Col. Trachta was transferred a year later, I bought his half for thirteen hundred dollars. Just prior to my discharge late in 1956 I sold it to a haber-

LOVE OF FLYING [173]

dasher in Maryland for three thousand dollars and four new suits. Thus I funded my formal legal education with cash and was able to enter it as a sartorial dandy.

During law school finances were meager, and I had to content myself with an occasional rental. I had established an investigating service specializing in trial preparation for trial lawyers. Whenever I could sell one of these worthies on the notion, I would take aerial photographs of auto accident scenes. This was accomplished by removing the pilot's window from a rented Cessna, swooping as low as I dared over the target, racking in a very steep bank, poking my camera out the window for four or five very fast shots, then grabbing the controls again to recover. Not a scientific method, but it did enable me to peddle the pictures for twelve bucks apiece.

The first few years of law practice in the early sixties provided lots of interesting and highly publicized cases, but not many substantial fees. I had to learn gradually that although my landlord was pleased to have a famous tenant, he did not regard press clippings as legal tender. Airplane sellers had the same unreasonable attitude.

By late 1965 I was able to forego renting and buy a new airplane, a Cessna 172. I loaded it with instruments and radios and began to fly myself on all but the longest trips; my practice was beginning to spread through a number of states in the eastern half of the country. In May 1966 my Cessna was sitting at the fuel pump when some genius whose battery was dead decided to hand-prop *his* Cessna with no one sitting at the controls. Unaware that the throttle was wide open, he spun the prop with a mighty twist. The engine caught, then roared, and six seconds later the back half of my airplane was a mass of shredded aluminum.

I was in a pickle. The extensive repairs required would take several months. During the summer I had to prepare for trial the cases of Dr. Sam Sheppard, Dr. Carl Coppolino, and Albert DeSalvo, the Boston Strangler. The airline pilots were on the verge of a strike, and I had a lot of ground to cover.

Inspecting the damage with me, and shaking his head, was my great and good friend Horatio "Mac" MacNamara. He had been renting airplanes to me for nearly ten years and had sold me what was now a wreck. I lamented my situation.

Suddenly he snapped his fingers, winked, and said, "Son, have I got a deal for you!"

And indeed he did. Parked in a corner of the field was a well-used Cessna 310, a fast, twin-engined airplane. It had lots of instruments and an autopilot and could be bought for *no money down*. The finance company that had repossessed it, Mac thought, would be content just to have someone pick up the payments.

Ten days and a dozen signatures later I was winging my way to California. The airlines were all but closed due to the strike, and the country was getting its first good look at the value of private air travel. The air charter companies were booming, airplane sales were booming, and I was all over the country.

In November 1966 Sam Sheppard was acquitted. The following month Dr. Coppolino was acquitted. Interviews in national magazines began to proliferate. *Life* magazine did a pictorial and showed my much traveled twin Cessna, now with new paint, interior, and engines. In the text of the article I was quoted as saying, "What I really need to cover all this ground is a Lear Jet. . . ."

In early 1967, when the article was published, the dream airplane of nearly every pilot in the world was Bill Lear's marvelous hot rod. With a basic design taken from the Swiss P-16 fighter, and against most industry predictions, Bill Lear, the aerial pioneer, had stood the experts on their heads once again. His small, powerful, swift, and incredibly sexy-looking little business-jet had captured the market place handily, and the imagination of every red-blooded airman as well. Frank Sinatra bought one, Henry Timken (of rollerbearing fame) bought one, Arnold Palmer and Arthur Godfrey learned to fly one, Jack Nicklaus rode around in one. By 1967 over a hundred had been built and sold at about half a million dollars a copy.

My remark had been little more than an offhand wisecrack, for the cost of a Lear Jet and my pocketbook were a helluva distance from one another. But when I was at the North Star Inn in Minneapolis, defending another doctor charged with murdering his wife, and a man named Ed Chandler called from Lear Jet to say, "I understand you might be interested in one of our airplanes. . . ." I did nothing to deter him from bringing it along.

Shortly before that, I had rented one for a couple of

weeks from Pat Haggerty, president of Texas Instruments, and had immediately fallen in love with the machine. When Ed Chandler and Lear Chief Pilot Hank Beard arrived to pick me up for a demonstration ride to LaGuardia, I hopped into the left seat like a man who had been born there. Serial Number 50A, once Bill Lear's private ship and now a factory demonstrator, was gleaming white with green trim. When I zoomed off the runway and haughtily checked in with air traffic control as "Lear 808LJ" my senses were tingling. I somehow had the feeling (not the funds, but the feeling) that possibly—just possibly—this beautiful, magical machine might be mine.

When we were level and cruising eight miles above the earth, Hank proceeded to show me that a Lear Jet with a single minigun could probably take on a squadron of Sabre Jets. I left Hank at the controls and stepped back into the cabin to talk with Chandler. My heart was beating rapidly as I waited for the cruncher—the number, the bucketful of dollars that would buy the airplane.

"Of course," Chandler began, "the airframe has about four hundred hours on it and the engines two hundred—we could make an allowance for that."

I nodded anxiously.

"In addition," said Ed, "ever since I read that article in *Life* I've been toying with an idea. We've had a little problem in the last few months with sales due to some tough publicity. There's been a rash of accidents lately and the airplane's getting the blame. Actually, it's just one pilot mistake after another. A lot of these pilots the FAA is checking out just don't get used to this much speed and power and the airplane runs away with them. But we're getting the blame. I thought that with all the attention you seem to get from the press, as an owner-pilot with a lot of jet time, you might be able to do us some good. Know what I mean?"

"Sure, absolutely," I replied. "Under such an arrangement, what price would you put on this machine?"

"I think," said Chandler slowly and with a smile, "that we could do a deal for about three hundred seventy-five."

My money-finding antennae began to vibrate. Between advances on a new book, the legal fee Sam Sheppard was about to pay, and some checking accounts I thought I could get about seventy-five thousand together within thirty days. A lender might be willing to give me the balance. "I think," said I, "that we can do that deal."

And we did. On March 29, 1967, I left Wichita with 808LJ and an FAA type-rating in my pocket. It was, in many senses, an entry into a whole new world.

I have many fond memories of my days as a Lear Jet driver. One of the fondest happened just a few miles from my home in Marshfield, Massachusetts.

August 1968 was a delightful month. The sun shone in New England, and particularly at Martha's Vineyard, a beautiful resort island just a few miles off the south coast of Massachusetts. Remi Brooke, the oldest daughter of Massachusetts' Senator Edward Brooke, was to be married on a Saturday. Mutual friends of mine and the Senator's had offered my Lear Jet to Remi as a wedding present with me as her chief pilot. She was to have it for her honeymoon. I liked and admired the Senator and didn't mind donating my time and my airplane.

I had been doing a certain amount of charter work for Massachusetts Governor John Volpe, who was campaigning hard for Richard Nixon. Rumor had it that Mr. Volpe would be designated Secretary of Transportation if Mr. Nixon won, and I delighted in having the opportunity to introduce him to air traffic control and its problems.

The day before the ceremony, an aide called and asked if I were attending the Brooke wedding on Martha's Vineyard. I said that I was. Could the governor ride down with me? Certainly.

The weather was perfect, the nuptials went off very nicely, and Governor Volpe extended his greetings to all. When he had finished, he opined that the reception might go on for some time, whereas he had some business to take care of at Otis Air Force Base, eighteen miles away across the water. Would I run him over? I would.

I climbed in the left seat, flipped the battery switch, and examined the fuel gauge. I shot a glance at Hugh Walker, my cocaptain, who was sitting to my right. We had about a thousand pounds of fuel on board. It would be enough.

Among the many marvels that combine to make up the Lear Jet, its tremendous power must be ranked first. As we taxied out, our gross weight was well under eight thousand pounds. The combined thrust of the two GE engines was fifty-seven hundred pounds. Had we been a little lighter, we could have climbed vertically.

The landing at Otis was routine. We glad-handed with a few generals and colonels who were on hand to meet the governor, then taxied back out for takeoff. The fuel gauge showed seven hundred pounds. The tower operator, who had no other airplanes to worry about, was in a chatty mood.

"I understand," said he, "that that Lear Jet is a pretty hot airplane."

"For a civilian airplane," I replied, "it is very hot indeed."

"What we have down here," he went on, "are airplanes with the afterburner. With a burner lit we can go right on by you."

"I'm sure you can," I said, annoyed at the affront to my precious Lear Jet. "But no civilian would be foolish enough to waste fuel on an afterburner. Now then, request permission for a short field takeoff."

"Short field takeoff approved," said the controller generously. He did not know that I was terribly light on fuel.

I positioned the Lear at the very end of the runway, wound the turbines up to more than 100 percent power until I thought they would scream out of their pods, and stood on the brakes. When I was convinced that the engines had no more to give, I released the brakes and rocketed into a moderate breeze. After a ground run of about eight hundred feet, I pulled the nose about forty degrees above the horizon, lifted the landing gear, and left the flaps down. We crossed the field boundary at nine thousand feet.

The tower controller, craning his neck to keep us in sight through his tinted glass, tried not to sound impressed. "That's not half bad for a civilian airplane," he drawled. "What do you boys do for an encore?"

Trying to sound just as casual I keyed the microphone and said, "Why son, we light the other engine."

* * *

Even if flying itself were less than a joy, the superb convenience involved would be worth the price. Great though our commercial airlines are, their shortcomings are severe. The need to arrive early, stand in endless lines for tickets and searches, and sit in a cramped seat with an elbow to either side vying for an armrest—these are annoyances most people would gladly do without. The flight may or may not leave on schedule, and it may not leave at all. Then again it may take off into the edge of a thunderstorm when it could well have waited thirty minutes and found much smoother air.

In peak travel seasons you had better not miss the flight you have ticketed, or you may be sitting standby in a packed terminal for hours or even days. And if there is a connection somewhere along the way—say at Chicago's very busy O'Hare Airport—and your flight is late getting in and the connection is missed, the rest of the ticket may be all but worthless if later flights are filled.

To a conservative traveler who cannot afford to miss his appointment, all of these "ifs" frequently mean that he must leave a day early and spend an extra night away from home. His productivity and his leisure time are both sharply penalized by the hours which must be wasted.

Finally, there are simply too many places where commercial carriers do not go, which necessitates a long drive in a rental car (if one is available) before reaching the ultimate destination. And yet there are few places in the United States which do not have a small field within ten or fifteen miles. In the long run it comes down to the difference in mobility and convenience between bus service and a private automobile, and there are few businessmen—other than those who work regular hours in the same office, day in and day out—who would find life tolerable if they were forced to get about by bus.

I had been flying for about fifteen years before I "discovered" the greatest fun vehicle of them all—the helicopter. I had had very little interest in helicopters and not a great deal of trust in their safety, an attitude of partial disdain that is not uncommon among fixed-wing pilots. [The helicopter's blades move; an airplane's wings do not. They are "fixed," thus the designation.] But I became involved with a Boston group who had purchased the Brantly Helicopter Company from Lear Jet in Wichita and felt that I couldn't contribute very meaningfully unless I learned something about the beasts. Somewhat grudgingly I went back to school and strapped on a helicopter.

To my great surprise, after a few introductory hours

of slicing up more than my share of sky as I learned to control the ship, I loved every minute of what was opening up as a whole new world in the air. Except for its limited cruising speed, the helicopter could do everything the airplane could do and a lot more. With its ability to hover motionless over a spot, fly backward or to either side, and land or take off practically anywhere, it offered a sense of freedom of dimension no fixed-wing pilot has ever felt.

And my misgivings as to the safety of the vehicle were very quickly allayed. I found it no more difficult to fly than an airplane and in no sense uncomfortable, but I did learn that with an engine failure it's a far better place to be.

A single-engine airplane that loses one engine is going to crash. Period. If he *happens* to be very close to an airport, or an expanse of smooth hard ground, a skillful pilot may "dead stick" to a landing without damaging the ship. But the odds against such luck are heavy and at night it's pretty hard to guess what's below you.

The helicopter is much, much better off. Because of its ability to autorotate—that is, the blades or "wings" keep turning even with the engine off—it can land safely and at a full stop in any reasonably level pea patch. As dependable as our aircraft engines are—and they are very dependable —it is comforting to know that in a helicopter you can get along nicely should the engine decide to quit.

Even before I had gotten my rating, I bought one of the little Brantlys and began to experience the sheer fun of flying all over again. I visited every acquaintance who had a yard big enough to land in, and as soon as I was fully licensed I repeated the visits and took them all for rides. For those who seemed apprehensive, I would immediately climb to five hundred feet over a clear area, chop the power back to idle, and autorotate to the ground. After a large grin of disbelief my passenger's apprehension would usually disappear.

Since those days seven years ago, I have practically stopped driving in favor of whirlybird transportation. Built onto the end of my home is a hangar with a landing pad set on tracks, which can be activated by the kind of radio controls commonly used on garage doors. On a rainy day it is possible to roll the helicopter out and take off, or land and roll her in, without ever leaving the cockpit.

In many senses, helicopters represent the best of both

worlds. They fly above the frequently congested ground traffic and below the sometimes congested airplane traffic, subject to very little control from anyone. Because they go from point to point in virtually a straight line, they are extremely predictable. A very nice way to waste less time.

For example, my home is situated some thirty-five miles south of Boston in a quiet residential community on the coast. The road in between is a superhighway, and in the dead of night, forty to forty-five minutes is ample time to drive the distance. But part of that superhighway is Boston's infamous Southeast Expressway, which many have called the world's largest parking lot. At rush hour, that is often a fair description. To make matters worse, the entire highway is also the main artery from Boston to Cape Cod, and on Friday and Sunday nights in the summertime it is packed. One accident can cause a snarl twenty miles long in minutes.

All of which adds up to a great deal of anxiety for the commuter. In order to get to town at a specific time to keep an appointment or a court appearance, how much time should I allow to drive thirty-five miles? The answer: anywhere from fifty minutes to two and one-half hours, depending on my gambling spirit that day. When traffic is heavy, there is no way to predict a jam until you are in the middle of it, and then there is no way out.

The helicopter simplifies things considerably. Cruising at one hundred miles an hour, the twenty-eight air miles take just seventeen minutes—and it always takes seventeen minutes, regardless of what's happening on the road or in approach control at the airport.

Even traveling from my home to downtown New York (there are several convenient heliports on Manhattan Island) the helicopter is more than a match for other forms of combined transportation. The hundred and eighty mile trip takes an hour and fifty minutes, door to door. Sure, an airplane can go from Boston to LaGuardia Airport in thirty minutes if there are no delays, but that is only takeoff to landing. Add in the start-up and taxi times, and the need to take a cab at both ends, and the total is very close, *if* there are no traffic delays in the air or on the road. If there are, it is no contest at all. Most important, I know when I will be arriving—other travelers are only gamblers, betting on the system.

As circumstances developed, nearly five years ago I made

the transition from helicopter buff to helicopter manufacturer, a step that even the most enthusiastic aviators are seldom likely to take. It happened this way.

Colin Gillis, then one of the senior associates in my law office, had been talking to a financier in Michigan who was convinced that helicopters had a great future and wanted to put some money into the Brantly, a fine little machine. One day just before Christmas, 1970, Gillis buzzed me on the intercom.

"Lee," he said, "you ever hear of a helicopter called Engstrom?"

"Sure," I replied, "only it's pronounced Enstrom, without the 'g.' They shut down early this year, which is good for Brantly because an Enstrom is three times the machine. Undercapitalized, I guess. I saw one two years ago at the Reading Air Show. Beautiful machine. I thought about buying one, in fact. Why do you ask?"

"It's these money people in Michigan," Gillis went on. "They just called and said they can buy the entire Enstrom Company, or 95 percent of it, at a right price. Wanted to know what you thought of the machine, and if you would take over the company and run it to get it back off the ground. They'll pay a salary and a hunk of stock."

I whistled softly. I knew something of the Enstrom Company. Its genealogy was typical of early aircraft companies —say in the late thirties—but in modern times was an anachronism. Born in the soul and imagination of a Michigan mining engineer named Rudy Enstrom and supported by some money and much enthusiasm by the border communities of Menominee, Michigan, and Marinette, Wisconsin, Enstrom went through more than five years of agony before being certified in 1965. By then almost all of the stock had been sold to the public. Money was terribly tight. Sales were few. The industry had grave doubts that despite its superior features, Enstrom could survive without a military contract, which it did not have. The company wobbled along.

Then in 1967 the giant Purex Corporation got a good look at the machine, calculated that helicopters were about to bust loose as a growth industry, and made a tender offer to the Enstrom stockholders. They picked up almost 95 percent of the shares outstanding and began to pour cash into the company as if they were printing it. Regional sales outlets were opened. A turbine power plant was installed and flown. Great happenings were predicted.

In February 1970, with less than fifty ships sold, Purex closed down the production line, released all but five employees (to keep the books, and to supply spare parts for ships in the field), and announced that Enstrom was for sale. They soon found that a drastic mistake had been made. Had they offered the company to buyers *before* closing it down, I think they would have found many takers. By the time the Michigan money people had started bidding, the price had fallen to a mere song.

I of course had no right to consider taking on the responsibility, on a part-time basis, of getting a shutdown aircraft manufacturing company back onstream. I was too damned busy with a multitude of other things. But the temptation was immediate and intense.

I had long envied the pioneers of aviation who had started their companies, fired them with imagination, drive, and a refusal to quit, and made them survive. Howard Piper, Clyde Cessna, Walter Beech, Donald Douglas, Bill Lear, Larry Bell, Stanley Hiller, Charles Kaman, Igor Sikorsky, Frank Piasecki—these were men who made it happen, whose names could be found on the manufacturer's tags of thousands of aircraft. Their day had come and gone, and pioneering was past. But if a fellow could resurrect Enstrom and make it fly? Well, history might find a spot on the second team. I had always loved airplanes and now I loved helicopters—and about the time involved, well, every man is entitled to a hobby, even a very serious hobby.

"Colin," I said, "I think I had best talk to these people. Get them on the line."

Two days later Lear Jet 808LJ touched down near Detroit and picked up the money people. Purex executives were flying in from California to close the sale that very day. They were anxious to get home the following day, which was Christmas Eve.

I taxied out and was cleared to the Menominee County Airport. Andy Crane, flying the right seat, told me quietly that the destination weather was up and down with snow showers, that no precision approach was available, and that the approach that was available led to a runway too short to land on unless it was absolutely dry. I nodded and said nothing to the money people in back. We could always go into Green Bay, Wisconsin, fifty miles away and rent a car to drive through the snow.

Forty minutes later we were descending over the western shore of Lake Michigan, talking to Green Bay Approach. It was snowing in Menominee, but there might be room. We were cleared to give it a try. As I descended from the final approach fix, I flew the gauges and Andy looked for ground. I was about to give up when he said, quite calmly, "I've got it. The short runway—no, there's the long one. We'll have to circle and hope to line up with the approach end. It's snowing pretty good."

I wasn't feeling reckless, and wouldn't have been if we were about to buy Boeing, but I was what you'd call a highly motivated pilot. I bent the Lear around as if I were flying a check ride with an FAA Flunk 'em Fanatic, and everything worked. As we came over the threshold I chopped the throttles, held the nose up, and let the three inches of soft snow on the runway slow the Lear. It was an unintentionally dramatic entrance to a small community that was to become my second home.

The papers were signed at the chamber of commerce building. The Purex people had never, in two years of ownership, even visited the Enstrom factory at the airport. They weren't about to visit now.

The following morning I toured the factory and flew an Enstrom for the first time. It was a dream. A dozen sister ships were sitting on the floor in final assembly, crying to be finished and adopted. The plant was as clean as a whistle —you could eat off the floor—and the workmanship in each of the ships was enough to make a flier's heart glow. All of the rivets seemed to have been set in place by people who cared much more that they be driven cleanly than about the quitting whistle. The employees still on the payroll were direct, straightforward, and trustworthy. The former employees who hoped soon to be back on the payroll were of the same stock. The community was cautious—their one prior experience with an "outsider" owner had flopped —but the support was there. I was feeling very good about the whole deal.

We all made it home for Christmas Eve. I felt as if I were six years old again, beseiged with visions of sugar plums, yet armed with the foreknowledge that Santa was in fact bringing me a whole company to run that made aircraft, to say nothing of a new demonstrator for the president to fly.

Several weeks—and snarls—later (the people from Michigan backed out of the deal) I bought just a little less than 95 percent of the Enstrom Helicopter Corporation.

We went back into business. We had to fight rumors that the clever lawyer from Boston had bought the company simply to liquidate it at a handsome profit. We began an aviation battle that only an aviation nut would take on, and yet we placed Menominee, Michigan, on the international aviation map. As of this writing we are "certificated" in 33 foreign countries and have built and delivered 620 helicopters. We now turn them out at the rate of twelve every four weeks.

You have seen our ships in *Jaws*, *Rollerball*, and on television and in magazines. The community still supports Enstrom as if it were a natural child, and the employees have always lived up to that first impression I got on a snowy day two days before Christmas in 1970.

My mother flies in an Enstrom. My wife does too, and so do my three sons. My dad thinks it's the only way to go. My friends call it the Great Escape, and some think it may be as close as we will ever get to the Arabian Magic Carpet.

From my point of view, manufacturing has been an adventure, a source of continuing delight, from time to time a burden, and always and ever a challenge. I strongly suspect that when the day rolls round that I am tired of trying lawsuits, weary of jousting with prosecutors and judges as to where I must be and when, drained of the emotional energy involved in helping solve other people's problems, I shall retire to that which I really like to do best: building helicopters, watching over them like a hawk as they come off the line, and flying—even if only for a few minutes—every single one.

2

The World of General Aviation

In the midst of the fuel crisis several years back, President Nixon came up with one of his least-inspired ideas by way of a partial solution: he suggested that fuel allocation for general aviation (by which he apparently meant everyone but the airlines) be cut back 40 percent. For the first time, at least in my experience, the usually diverse world of general aviation rose up with a single voice in opposition.

I guess the President thought his idea would appeal to the vast number of voters who could not afford an airplane and therefore might enjoy the discomfort of those who could.

Perhaps he thought, or was advised, that there would be little chance of the traditionally splintered mix known as general aviation—which runs the gamut from airplane manufacturers to the guy who owns a one-third interest in a battered Piper cub—ever agreeing on anything. Yet for once they did agree. Fortunately, the manufacturers had just formed an organization, GAMA, the General Aviation Manufacturers Association. They were joined in opposing the President's plan by the already existing AOPA, the Aircraft Owners and Pilots Association, the National Pilots Association, and a number of other related groups.

Nixon should have foreseen that general aviation would have a strong ally in the Congress itself, where many members rely on private air transportation. Not only are there a good number of senators and congressmen who are, or were, pilots themselves, but few politicians relished the thought of suddenly having to fly commercial, where fellow passengers have been known to plunk down in the next seat and deliver their polemic for the day, or say something to the effect of, "What did you do for me today, Senator?" They could see that life under the Nixon plan would soon become intolerable, especially on the campaign trail.

The White House idea was discarded, but, as I mentioned, it was about the only time in my memory that all the parts of general aviation coalesced. It is hard to expect much unity from a group that contains at one end of the spectrum (at least in theory) the sophisticated and highly-trained quasi-businessman/pilot, and on the other the sportflyer who uses his machine only when the sky is blue, and then for only one hundred hours a year. Their needs and interests are so different that it is hard to get them to consolidate their point of view so that a single position, acceptable to both, can be taken. Another problem is that general aviation has traditionally been the whipping boy for almost every ill in the system. When major airports are choked up because the air carriers are overscheduling, the blame is always laid at the door of general aviation, even though the figures don't back up the complaint.

There is even some difficulty in trying to define the term general aviation. Too often it has been used as a catchall to mean everything that is not an air carrier or a military plane. Even the view taken by the FAA is probably too broad, because it includes—which in my opinion it should not—those sophisticated private and corporate-business aircraft that are equipped in every respect like airliners (and can outperform many of them) that are flown by experienced professional pilots of the caliber you would expect to see wearing four stripes for a major airline.

My own Rockwell Commander 690B would be a good example. We probably have more and better avionics than an awful lot of the DC-9's and 727's that are flying the airways. In fact, I've had cockpit crews tell me that. It's pressurized, sophisticated, fast, and has plenty of backup systems. It can fly anywhere an air carrier can go, except for over-water flights that are beyond its range because it doesn't hold as much fuel. Actually, it can fly many places where the big air carriers cannot.

To me, general aviation, classically, is the group that flies for fun, to drill holes in the Sunday afternoon sky, to go on trips, or perhaps just to visit the Coke machine at the airport fifty miles away and shoot the breeze with someone different for a change. Also included are those who use their planes for occasional, but not incessant, business purposes. Within this group there is, as might be expected, a wide variance of pilot skills and understanding of the entire system.

In order to get a better grasp of the concept of general aviation, let's categorize some of the pilots it includes.

The Student Pilot

Say a person has no other goal than simply flying himself around. He (or she) can become a student pilot, attend ground school, and begin to get in some air time. His student ticket is good for two years, but if he doesn't upgrade to private pilot, he has to take another check to remain a student. I've known some student pilots who stayed in that category for ten to fifteen years; they had no desire to carry anyone else along with them, and saw no reason why they should progress further. But that's not the normal situation. Most student pilots stay in that category only as long as they have to, which depends on skill, time, and money [The average cost for getting a private pilot's license is around $1,200]. According to AOPA figures, as of January 1, 1978, there were 203,510 student pilots in the United States.

The Private Pilot

With 327,424, this category is by far the largest in all of aviation, making the private pilot certificate the most common.

The private pilot can carry anybody he likes with him, but he may not accept money for it. Occasionally, the rules are bent to allow for sharing of expenses on joint ventures, even including business trips. But, in theory, if there is compensation of any form, the private pilot's license is not enough.

Like the student, the private ticket holder must take an FAA physical in order to obtain his third-class medical certificate. In 1977 the FAA issued 54,657 private pilot licenses.

The Instrument Rating

This is generally regarded as the big hurdle, the step that separates the men from the boys, the Sunday pilots from the serious air travelers. Anyone with a Private or higher rating is eligible. Those without it are called Blue Card Holders. Supposedly, VFR pilots carry a blue card with a hole in the center which bears the following legend: "When you hold this card up to the sky and it is all blue, you may go." Because of its significant position in the airman's world, the instrument rating is discussed in some detail later on in this chapter.

The Commercial Pilot

This next step up the line (like the private and instrument ratings) requires a pilot to pass a written FAA test and a rather lengthy flight check, also given by the government, to test the precision of his flying and the extent of his understanding and ability to interpret all the weather information that is available to pilots. It represents an across-the-board upgrading of all the skills required of the private pilot. According to the latest AOPA figures, there are 188,763 pilots who hold a commercial rating.

For years it was possible to be a commercial pilot without an instrument rating, and many were. Although it is still *possible* to be so rated, it is hardly worthwhile. As of November 1, 1974, the FAA changed the rules to limit non-instrument commercial pilots flying for hire to flights of no more than fifty miles in daylight only. Therefore, except for those who wish to give scenic rides around the local area on fair-weather days, all commercial pilots must now be instrument rated.

The Air Transport Pilot

Although this highest of ratings is required only of those who are flying as captain for an airline, it is held by a great many people (50,149 in 1978), including some who do not require it, but do so probably for prestige if nothing else. In essence, it is simply a super-tough version of the commercial-instrument rating, prefaced by a brutal written examination and capped with an exhaustive and challenging check ride, described by Captain Don in an earlier chapter.

The Flight Instructor

The flight instructor rating is available to those who hold a commercial-instrument ticket and can pass both written and flight examinations specifically testing instructional ability. These ratings are specially designated as to the type of instruction which the instructor is licensed to give (multi-engine, instrument, seaplane, glider, helicopter, and so on). To keep the rating current, an instructor must teach to completion every two years a total of ten students in *each* category in which he is licensed, and if he fails to do so as to any category, he must take a new test to retain his authority to teach further.

Category Ratings

Although most fresh student and private licenses are issued as "Airplane—Single-Engine Land," this is not a requisite. Some pilots have gone directly into multi-engine aircraft, some have started with seaplanes, some can fly only helicopters, some only gliders. More typically, these other category ratings are the product of a follow-on effort by pilots who have learned the basic skills in a single-engine airplane with wheels. In each category, the applicant must show his skill in the aircraft, especially in those areas where its operation varies from that of a standard airplane.

Type Ratings

For every airplane that weighs more than 12,500 pounds, and for every airplane powered by pure jet engines regardless of weight, a pilot must take written and flight examinations for that particular plane before he is "typed" and therefore allowed to fly it as pilot-in-command. Because management of aircraft of this size and complexity requires substantial skills, and because virtually all of these ships are likely to be flown in all weather conditions, the FAA is usually very fussy about issuing a type rating, and its examiners are apt to be rather exacting in satisfying themselves that the applicant has both the confidence and dexterity to control the airplane through a nerve-wracking series of emergencies.

License Limitations

One might get the impression from glancing at the above qualifications criteria that only perfect specimens (with straight teeth and crooked smiles) are allowed to apply for flight ratings. This is not true by any means. The FAA is in fact very good about trying to help those who suffer from one disability or another to earn pilot status, provided no serious hazard to safe flight must be overlooked in the process.

As an example, one of our Enstrom dealers in the Midwest flies a helicopter well and on a daily basis, and he has a wooden leg and one eye. A man with no legs was once licensed to fly an Ercoupe, a little postwar trainer that wouldn't spin and required no rudder pedals. All sorts of people with eye trouble are licensed, with the limitation that "corrective glasses must be worn." People who have suffered heart attacks have won their right to fly again, although only after the most careful scrutiny. This caution is certainly warranted, since the recurrence of heart attacks is extremely difficult to predict, as I once learned to my profound embarrassment.

A delightful gentleman named Frank Pristow from Johnstown, Pennsylvania, had had what the FAA believed to be a heart attack, and the FAA revoked his medical certificate. Frank felt that if he couldn't fly his Cessna 310 around, life was hardly worth living, and so he set out to prove that the seizure he had suffered was indigestion, not cardiac trouble. He hired me to help him.

I sent Frank to a hospital in Cleveland that specialized in anginacinematography, a medical technique wherein pictures are taken of the *insides* of one's veins and arteries to determine the amount of plaque which is clogging them up. The results were encouraging. True, there was some plaque buildup, but nothing abnormal for a man of fifty-three. We obtained a hearing before a National Transportation Safety Board examiner and went to bat.

I put in the scientific evidence described above, and the FAA took its turn in opposition. When the FAA medical expert took the stand and defended the administrator's position in refusing to give back Frank's medical certificate, I got him to admit that the test results were within the range of "normalcy" for a 53-year-old male. I then asked:

"And now, doctor, how many airline captains whose blood vessels contain 'normal deposits of plaque' and who are fifty-three years of age or over are now flying the line?"

The doctor looked startled. He didn't know the answer in numbers, but he allowed as how there were a good many, probably a majority. I could see a slight smile playing along the examiner's face.

The evidence closed, and FAA counsel told me, "I've worked with this guy before. You're going to win. We don't want to ground half the airlines because some crazy private pilot wants to kill himself."

The examiner said, "Gentlemen, I believe I have sufficient evidence before me to make a decision. Please have your briefs to me within a week, and you will have a judgment very swiftly." He may not have winked at me, but I thought he did.

I worked all weekend to prepare a brief. When it was done, I thought I had demonstrated absolutely that Frank Pristow had never had a heart attack and most certainly never would. I mailed it to the examiner shortly before noon on Monday.

Less than one hour later I received a call from my associate counsel in Johnstown, Cal Abood. "Bad news," he said. "Frank Pristow died an hour ago of a heart attack."

"You're kidding," I groaned.

"I am not," said Cal.

It was too late to retrieve my brief from the Postal Service. Whatever his disposition had been, the examiner would have a chuckle when he read it. A little egg on the face is every man's risk, but at that point I felt covered from head to toe.

Frank, however, had apparently been as beloved by his community as he was ill. Should you chance to stop for fuel or a Coke or just to chew the fat at the lovely airport facility in Johnstown, Pennsylvania, you will enjoy your stay, however brief, at what has since been officially named Frank Pristow Field.

* * *

In the progress of a pilot, however, the single most distinguishing achievement is undoubtedly the instrument rating. More people stop short of this rather expensive and drawn-out bit of education than they do at any other level. Actually this isn't really so bad, because it means there are a great number of pilots who are unable to join the system on a cloudy day when the air traffic controllers are loaded with work, which makes for a rather happy natural balance to the whole affair. Pilots without an instrument rating are restricted to flight free from clouds, and in most circumstances, always in contact with the ground.

The instrument rating requires an entirely different discipline, as does the initial indoctrination into flight itself. The big transition in going into instrument work is that the pilot learns for the first time that he must rely on the impressions he gets from certain senses and completely reject others.

We humans rely for our total sense of balance on a nervous system found in the inner ear; someone with a good sense of balance can walk a tightrope, but one whose system doesn't function quite as well might only be able to negotiate his way along a seawall, providing it is a foot or more wide. This sense of balance is informative when one is flying normal maneuvers, because if an airplane gets into a slip or a skid, one can feel it in "the seat of the pants."

However, when shifting to instrument flight, all of the sensations of the inner ear must be disregarded. In the first place, they are deceptive. One can be perfectly straight and level and still swear that he is in a turn; or he can be in a 30-degree banked turn and believe that he is straight and level if he doesn't look at his instruments.

So, it is truly a disciplinary feat to gradually convince yourself that any time your eyes and ears disagree with the information they are presenting you, the eyes must control. And the balance information must be rejected.

For pilots who don't go through this transition, but nonetheless get into the clouds and fly on instruments, disorientation and subsequent accidents (where either the plane goes out of control at sufficiently high speeds to tear itself apart, or simply pops out of the clouds in a spin too low for the pilot to get his senses back and recover) are all too common.

I have to say, at the risk of sounding like an old bold pilot, that things have changed mightily since 1953 when I went through the Navy's all-weather school. Then it was a rather complex art demanding a certain amount of talent to fly instruments well. But there have been tremendous advances in the presentation of the information that the pilot now sees, and the consolidation of a lot of little bits of information that used to have to be derived from a rapid scan of some very different looking instruments, all of them presenting very different things you had to know.

The instrument panel of the early fifties was equipped with such standard items as an airspeed indicator; a rate of climb indicator to show whether or not you are level, descending, or going up; an altimeter to show your exact altitude (or as close to exact as possible, because there's always a little lag in the barometric altimeter); a turn and bank instrument to show whether you are in fact in a turn and whether or not the aircraft is trimmed properly in balanced flight; a compass or directional gyroscope which would show the heading of the airplane; and then an attitude indicator, which is a small airplane symbol fixed to the instrument panel and facing a horizon which is fastened to a gyroscope, so that no matter how much the airplane might contort around the instrument by banking, pitching, gliding, or climbing, the horizon always stays level, as a reflection of the natural horizon which may be obscured by the clouds.

In order to fly with precision, you simply had to keep your eyes going constantly—and it could be very hard work.

Today's planes are, or can be, equipped with an array of instruments that all but take the work out of it. Some of this equipment is rather expensive for the owners and operators of light airplanes (and there is an argument currently raging with the FAA for having required certain costly additions) but no one quarrels with the sophistication of these devices. One of the most advanced is the "flight director," which performs several tasks, almost like an avionics committee. Its main purpose is to give the pilot, as he studies a single instrument, as many different pieces of information as possible telling where his airplane is, and *where it ought to be going.* This latter feature, a product of the computer age which has been available for the last ten years, is the real work saver, for it saves the pilot a good deal of very pressured analysis and reasoning.

Before flight directors, six different instruments might combine to tell the pilot flying his final approach on an instrument landing system that he was (1) off to the left of the localizer beam, (2) below the glide slope beam, (3) descending more rapidly than the glide slope could accommodate, (4) too low above sea level, (5) too low above the ground, and (6) flying at a dangerously slow airspeed. After absorbing and analyzing all of these indications during his rapid instrument scan, the pilot would have to reason out corrective action and then take it, by (a) adding power, (b) banking to the right to change heading back toward the localizer, and (c) raising the nose to get back up to the glide slope.

But when the flight director is used, both the problem and the correction to be made are analyzed by the computer. The face of the instrument will have moving lines or bars called cues, and merely by following them the pilot will be led back to his rightful place in that piece of sky. In addition, during the final and critical phase of the approach, his altitude above the ground will be displayed, as will his ground speed.

To make matters simpler, if the flight director is coupled to a sophisticated autopilot (and most of them are) all the pilot need do is watch. The autopilot will make the necessary adjustments and the pilot need only control his airspeed with the throttles. On the latest models, the pilot dials in the desired airspeed, punches the button marked auto-throttle, and folds his arms. The system will put him right over the threshold of the runway right on the centerline at the correct speed. He need only flick the disengage button, flare, and land. Or, if the visibility is so low that he never sees the runway, he will push the go-around button and the aircraft will automatically lift its nose, add power, and begin a turn to follow the missed approach procedure. The pilot need only raise the wheels and flaps as soon as a satisfactory climb has been established.

Needless to say, these are wondrous devices, and when they work they can make instrument flying a piece of cake. Also needless to say, the FAA does not permit their use during an instrument flight check. The most marginal of pilots might get by with all of the goodies operating. Flight checks seek to determine who can fly when computers are not working and a few other things are going wrong to boot.

If you fly for a living, you will have an instrument rating

LOVE OF FLYING [195]

(unless you fly helicopters) or you won't be making a living for very long. But if you fly for pleasure or business or a combination of the two, you may earn or forego instrument qualifications as the mood may strike you. As I have said earlier, I think flying by the gauges is both challenging and fun, builds confidence, and enhances safety.

To all but the purely sporting types, I heartily recommend getting the rating, even if you have to chip away at it gradually over a considerable period of time. If ever you need your airplane to go from point A to point B and back again on any sort of schedule, your right and ability to go IFR will just about triple your chances of success. And if you don't have the rating and the weather squeezes down just a little bit, the temptation to try to slide between that cloud and that ridge becomes irresistible to the VFR pilot anxious to get home. Those squeezed down ridges have become "home," in a very final sense of the word, for a great many such pilots.

Short of a full rating, there are some types of training which, although abbreviated, are designed to enable a pilot to get out of the trouble he is in and return to the good skies he just left. For those who will not or cannot pursue a full instrument course, this training is of tremendous value.

Quite a number of years ago, the Aircraft Owners and Pilots Association, which represents all kinds of airmen from air transport pilots to those who only plan to learn to fly, instituted a course called the 180-Degree Rating. It was, and was intended to be, as simple as possible. If you have inadvertently flown into a rainstorm, snowstorm, or cloud and you are not instrument rated the safest possible move is to go back where you came from. This may sound like I am declaring the obvious, but to the uninitiated it is much easier said than done.

First, it requires an admission that the pilot has made a mistake, and that he is in trouble—two conditions that human beings are very reluctant to admit. Second, it requires that one who flies his plane by reference to the horizon and the feeling in the seat of his pants, disregard both and pay attention to his instruments, a discipline to which he is unaccustomed. Third, the pilot must remain in firm control, put panic aside, and make a gradual turn in reasonably level fashion until his heading has reversed itself and he is pointed back toward his own environment. Provided he has acted decisively and swiftly, his own comfortable world will return in a few moments and the adrenaline can return to normal levels at the same time.

Training of this sort may seem almost childish, but I am personally convinced that the 180-Degree Rating has saved a bucketful of lives. Now, fortunately, private pilots are given as part of their normal curriculum enough instrument familiarization to get out of *inadvertent* instrument difficulty. I suspect that AOPA had something to do with that most desirable flight training improvement.

Quite apart from instrument flight, there is another phase of general aviation—also initiated by AOPA—which to my mind is both interesting and a helluva good idea. It is called the Pinch-Hitter Course, and although designed primarily for private pilots' wives, or husbands, is a nice piece of insurance for anyone who flies in private aircraft but is not a pilot himself.

This course, and others of its type, are intended to accomplish only one purpose: to aid a passenger whose pilot suddenly becomes disabled in flight to get the aircraft back on the ground without injury to the occupants. If injury to the aircraft can be avoided, so much the better, but in these conditions any landing you can walk away from is a good landing. As a manufacturer selling aircraft to people who will use them at least part of the time for pleasure, flying their mates and friends around, I am a strong supporter of the pinch-hitter concept. We can always build a man a new helicopter, but we can't build him a new arm, leg, or family.

The primary objective in training one to pinch-hit is precisely the same as that involved in getting out of weather trouble or surviving an emergency: Don't panic! If the pilot suddenly gives a gasp and turns gray, take the controls —gently—and hold them steady. Know how to use the radio, switch to the international distress frequency (121.5 mhz, a number that should be burned into the right wrist of anyone who fools with aircraft), and start transmitting "MAYDAY!" Except in the most remote areas, someone will answer and soon.

A true pinch hitter will have had enough training to manage the aircraft well enough to get it down with a little help on the radio. But there are many instances of passengers who have never *touched* the controls making a safe landing after being "talked" down, sometimes by a controller on the ground who knows how to fly, sometimes by another pilot who can join up in formation and give detailed instruction, and sometimes by a combination of the two. Any familiarity and skill that a passenger can develop will be very helpful of course, but the bottom line is always the ability to communicate. If a nonpilot passenger suddenly finds himself forcibly in charge of an airplane, the absence of panic and the ability to use an aircraft radio will give him—or her—a fighting, and often good, chance to fly another day.

Many people who are grudging victims of my voluble enthusiasm for the world of flight are wont to ask, "Why do you think flying's such a big deal? Why should I spend the time or the money to learn? What's it going to do for me?" Curiously, these same people are apt to ask, "How come you don't use any notes when you give a speech? Why do you look so relaxed in a courtroom when the air is crackling with tension? Don't you quaver a bit when arguing before the United States Supreme Court? Aren't you nervous on the Johnny Carson show?"

In a very real sense, one set of questions provides the answers to the others. If there is one great asset with which to take on life, it has to be self-confidence. And I think that no activity known to man can build self-confidence more rapidly and more surely than making him an airman. I have the greatest respect for the justices of the United States Supreme Court, and meticulously prepare each case I take before them. But I am mindful that they are also men who don their trousers one leg at a time, and who do not currently come to the Supreme Court by shooting an instrument approach at minimums with a load of ice on the wings. As for Johnny Carson—well, he's never nervous, but then he's a pilot too.

I think it is fair to say that nearly half of the flight instructor's task is to gradually instill in his student the belief that he is the master of the airplane, rather than vice versa. Initially, this is done by demonstrating how gracefully the aircraft will react to even the most timid and gentle commands. In more advanced stages, that confidence is sharpened by making the airplane misbehave and convincing the student that he has the capacity to cope with that misbehavior. Ultimately, it is solidified by forcing the student to make all the decisions as if he were alone in the cockpit, with no big brother and no big daddy to call upon if something goes wrong. When this stage is reached, it is time to turn the airman loose. He has reached that degree of self-reliance where he is entitled to be called a pilot.

In my view, general aviation, as that mixed bag is called, is of great value to the United States. It is the antithesis of what this country too easily tends to become when it is not unified by some external threat: security conscious, riddled with me-tooism, flaccid, a population determined to depend on someone or something. Good fliers make good people, good leaders, and good citizens.

There is some notion that flying is for the world of the well-to-do. Any Sunday visit to a small airport will teach that this is simply not so. There will be scores of people on the student rolls who have to scrimp from a modest paycheck to afford an hour of instruction. There will be teenagers with stars in their eyes pumping gas and washing airplanes, trading their labors for an occasional hour aloft. And there will be a bunch of old but shiny airplanes owned by partnerships of blue-collar workers, costing the price of a decent used car, that are the pride of the ramp in their owners' eyes. In the flight office or the hangar, the part-time pilots will recount to each other their endless hairy tales, punctuating their stories with moving hands to show how the airplane flew, beginning as often as not with, "And there I was flat on my back at ten thousand feet. . . ." Plumber and banker, gas station attendant and doctor, they have a special language and they are a special breed—simply because they believe they are. After all, almost anyone can fly an airplane.

I once told a jury in Jacksonville, Florida, a most significant jury for in that case my client and his lawyer were wearing the same suit, that I was grateful to the United States government for two things in particular: under its auspices, I had been taught to fly and taught to try a lawsuit. I meant that, very sincerely. But to it I should add that I'm glad I was taught to fly airplanes first.

Recommendations

> Through the window above the kitchen sink,
> past the log pile stanched between two trees,
> past bare branches of dogwood, hickory,
> splashing against winter skeletons
> is the blare of your red coat and I freeze
> with the dishcloth in my hands, freeze
> with terror of the familiar.
> What do I know that I cannot know?
> > "CLEARED FOR APPROACH,
> > CLEARED FOR LANDING"
> > —*Ann Darr, 1975*

Ann Darr is a poet. She and her husband live in a suburb of Washington, D.C., but they have a small house in the Blue Ridge Mountains where they go for quiet relaxation in all seasons. They were in that wooded retreat on the morning of December 1, 1974, when TWA 514 labored through the nearby skies, much too low.

A good ear for sounds is an asset to a poet, but on that awful weather morning, Ann Darr reached back to an experience of some time ago, back to the second World War when she was a pilot who towed targets across an artillery range. The moment she heard the sound of the plane, she knew it was not where it should have been.

Within a short time, the premonition that had caused her to "freeze with the dishcloth in my hands, freeze with terror of the familiar" had become a cruel, shocking fact. A plane had crashed and ninety-two people were dead.

Six months later, Mrs. Darr finished a long poem, her personal catharsis; in it she deals with the question that all of us should consider, the question of blame:

> At eleven 09 on Sunday morning
> the plane droned low over our house.
> 20 seconds later it flew into Weather Mountain.
>
> I am not responsible. I have accurately
> figured that I could not have reached
> the telephone, called the field, alerted
> the crew. The moment of crash is exact.
> The speed of aircraft is recorded. Twenty
> seconds is not enough time to reach up
> my hand in the air and stop that flight.
> Now you know who I think I am. And
> guilty of everything.

On December 1, 1975, the Washington newspapers ran their traditional "one year ago today" story about the crash of 514. One of the papers decried the fact that scavengers were still poking around the area, despite the signs and the fences, looking for souvenirs. But the paper added to the ghoulishness by using two stark pictures, one of an article of clothing caught high up in a bare tree and another of a single charred tennis shoe near a blackened tree stump. I suppose that kind of coverage is to be expected. Perhaps it even does some good, for it certainly forces people to think. Whether they think about air safety is the question, though.

The papers could have saved their film and simply reprinted a few stark lines from the Darr poem:

> My life passed before me in the middle
> of the fog on Sunday morning
> and made a bee-line for the peaks.
> Flew into a mountain. With all
> aboard singing and dancing
> and the altimeter working, after all,
> and the flight path accurate
> and the hearing optimal. So what
> of the rain and the wind and the fog?
> Under that throttle is power, and with
> that power we flew into Weather Mountain.
>
> A head is hanging
> in a tree. Firemen hunt

> for bodies. I wonder about
> the fingerprints when
> the fingers are forsaken.

Harsh imagery? Of course. But also quite true. There is no saving grace in a fatal crash, unless it's supernatural. Ironically, December brought the end-of-the-year estimates that air travel had a better safety record in 1975 than in the previous year, which in no way consoles those who lost loved ones in the crash of 514.

The problem is that we are still reacting to air safety problems *after* the fact. Far too often the danger signs are read but no truly remedial action is taken until a tragedy shakes us out of our lethargy. The time to be concerned about air safety is always *now*.

As a nation, we can and should be proud that our airlines have the best safety record in the world. But that is no cause for a lack of diligence in the areas where improvements are needed. It is worth thinking about that in the modern jetliners, the pilot's single greatest problem is complacency.

As I've said before, my roots in aviation are rather long for someone in his early forties, and my role is at least threefold: pilot, manufacturer, and very frequent passenger. But when I step in someone else's airplane, strap myself in, and start flipping through the pages of a magazine, my chances for arriving safely are exactly the same as those of the guy across the aisle who's chewing his fingernails because he's never been off the ground before.

So, let me offer the following recommendations and suggestions for improved air safety.

The Certification Process

Despite the fact that the law does not presently favor the plaintiff who tries to include the government in his suit against a manufacturer for an accident caused by an unsafe—but certified—part, it may be time to question the logic and the fairness of this situation.

That, however, is not my main point. There have been some serious problems in recent years involving unsafe conditions that have cropped up *after* certification has been granted. Most notable, of course, is the tragic situation of the cargo doors on the DC-10, wherein it was learned that the FAA disregarded the recommendation of its own regional office for an Airworthiness Directive, and, instead, allowed the manufacturer to deal with the problem in its own fashion.

This kind of collusion between the government and the industry has to stop. I would suggest the agency construct some form of internal mechanism whereby the recommendation of an Airworthiness Directive from a regional office be given almost quasi-legal status within the agency. By that I mean, there should be some sort of appeal process so that the people in the field who made the recommendation (for the toughest action the agency can take) be allowed to present their case a second time before anyone at headquarters can summarily reverse (or subvert) them.

I feel strongly about this not solely because of the DC-10 tragedy, but for the simple reason that in many cases the people in the regional offices are really in tune with the situation, far more so than the bureaucrats in Washington. I know the hardship that can be caused when a company has to shut down its entire production line (there have been cases where a mandatory fix, which is what an AD calls for, has caused a company to go out of business) plus, on occasion, having to call in all the planes off the line. But that still does not justify the loss of a human life.

The FAA and the Industry

Problems relating to certification are in a sense the tip of the iceberg. There is a long and none too proud history of the FAA's coziness with the major aircraft manufacturers (just as there is between the CAB and the big airlines over the issues of rates and routing). John Shaffer was hardly the first FAA boss to play footsie with the industry, and unfortunately he will probably turn out not to have been the last. The pressure that the industry, through its lobby, is able to put on the FAA is too great to be overcome in a short time.

But there are steps that can be taken. Several years back, the Congress attempted to shed some light on the way it is "lobbied," by passing the Federal Regulation of Lobbying Act in 1946. It has turned out to be a rather toothless power, and various attempts to toughen it have not prevailed, but the idea is one that could well be helpful in bringing some fresh air into the corridors and backrooms at the FAA.

The Role of the FAA

Although I had run across the paradox before, writing this book vividly brought home to me that the FAA is probably alone among the government agencies in engendering both love and hate. Well, maybe love is too strong a word, but hate, all too often, seems to fit.

Those who praise the FAA do so because of the careful way in which so many of its faceless employees all around the country go about their jobs. These are the people who check airport equipment, check out pilots, check out flying schools, airplanes—you name it. In almost all cases they know the vital importance of their jobs and they take them seriously, without pulling rank.

On the other hand, the bureaucrats who run the high-level departments do not seem to come in for very much praise. They are accused of being primarily concerned with the agency's image, as opposed to the facts of various charges and criticism. And, as already mentioned, they are charged with being much too concerned with the point of view taken by the industry and the airlines.

The FAA's attitude toward the air traffic controllers has improved greatly, especially since the formative days of PATCO—but then it had nowhere to go but up. Even today that relationship is hardly a model for other areas of government. On a segment of CBS's *60 Minutes*, aired in early 1976, then Acting Administrator James Dow was questioned as to why the FAA would not verify what several of its Jacksonville controllers had told the reporter—that the military was using commercial airliners as "practice run targets," veering away at the last moment. His answer was a classic in evasionary tactics. Apparently the controllers are still having trouble getting their complaints heard upstairs.

A relationship of equal, if not greater, inportance is that between the FAA and the NTSB (National Transportation Safety Board). Although the aviation safety community is hardly agreed on this point, I believe that the time has come to give the NTSB the power to enforce its recommendations. And if it takes "independence" to accomplish that end, than I am for it. I for one am tired of seeing glowing stories in the general press about the tough recommendations made by the board, because I know how simple it is for those recommendations to be ignored. The NTSB has the expert investigators; it should also have the force to back up its findings.

There are some highly-respected people who disagree with this idea—that of an independent, powerful NTSB—on the grounds that the board is not large enough and therefore might make mistakes that the larger body would not. Somehow, that's a chance I'd be willing to take. I think this is a good point at which to mention that the NTSB (at least in the form that it then had) was calling for a ground proximity warning device as early as 1963. All planes were to have the device installed by December 1, 1975, but that date came and went and only two thirds to three quarters of the domestic fleet of airliners were carrying them. The Air Transport Association convinced the FAA that the airliners needed more time; the FAA agreed and gave them extensions ranging up to ten months. I wonder if the NTSB would have been as generous.

I do feel that the NTSB should be independent (if that is the way to give it enforcement power) but I am not as yet convinced that the FAA should be. The FAA may be part of the Department of Transportation, but it is by far the greatest part. Perhaps it would be best to leave it that way —provided the Secretary of Transportation is sufficiently strong. According to the "chart," as government employees are wont to say, the DOT boss is more important than the FAA boss. The latter works for the former, at least that's the way it looks on paper.

If the proper lines of force can be renewed and maintained, then I vote for the status quo. If not, then the time is at hand to cut the FAA loose and make it independent of the Department of Transportation. Some people say that is the way to provide for accountability. They may be right. At this point I'm just not sure. One thing that worries me is that I have never heard of a new government body, Department or Agency or whatever, that did not suddenly begin with the first breath of its new life to grow. I think the FAA is big enough already. I don't want it to grow; I just want it to grow up.

I would also recommend that the fine work begun by the House subcommittee that had jurisdiction over the FAA be picked up and expanded by its successor. One air safety expert, Jack Carroll (on loan from the NTSB to the Flight Safety Foundation, which he serves as president) said that he had never seen a congressional committee "pick up on and zero in on its subject so quickly" as the House group that made the FAA study. It would be a shame, and a great waste of time and money, if that study remains on the shelf. It raised the proper points; but we are still waiting for the answers.

I'm told that the new subcommittee, Transportation and Aeronautics, has some fine staff people (one is a former vice-president of Eastern Airlines) and has begun to hold hearings to acquaint itself with its new assignment. That is all well and good, but it should not forget that its predecessor left a report that is a wealth of information. It tells them where to go. They should follow the leads.

Finally, to repeat as a recommendation an opinion listed earlier: the FAA need no longer concern itself with the growth of aviation. Aviation has come so far that it can certainly take care of and promote itself. The prime and sole concern of the FAA should be air safety, and if it doesn't recognize the inevitableness of that fact, the Congress should.

The Role of the Airlines, Their Pilots, Crew Members, and Other Employees

Given the amount of governmental regulation that already exists, it is hard to expect that the airlines would take the initiative in implementing further safety-related changes. But it is not too much to ask, for, traditionally, many airlines have instituted such changes long before the government made them do so. (For example, Pan Am had ground proximity warning systems on their planes more than a year before they were required by the FAA, and certain Boeings were coming out of the factory with the device as standard equipment months prior to its becoming mandatory—if a buyer didn't want it, he had to pay to have it removed.) Such concern for the safety of the flying public is by no means rare.

There are areas, however, where the airlines could do a lot better. I have not really gone into it in this book—because there is no way of getting accurate statistics—but there are times when the profit-motive of the airline and the personal motivation of the pilot combine to create a situation where both the company and the man wink at the safety regs.

I'm talking about the type of situation that is mentioned in every book or article on air safety ever written by an airline captain: the situation where the weather has closed in and threatens to cause either a diversion or a cancellation of a flight. In these cases, many pilots will not serve the schedule, thereby angering some passengers and probably displeasing the company. Unfortunately, there are almost always some captains who will do just what the company wants—even if it means cutting the safety edge very fine.

I'm not suggesting that these pilots are knowing fools; they are undoubtedly convinced that *they*, as opposed to other pilots, can fly that approach or make that borderline takeoff. It's too bad, but they have to realize that the company will be pleased by their decision to keep the schedule on time and in place. One can only hope they are as good as they think they are. But an airline should never encourage such pilots.

This particular problem is not limited to flights in bad weather. It can be seen in regard to the condition of the planes. An Eastern stewardess once told me that she—and many other flight attendants—has been put in the awkward spot of having to decide whether or not to fly on a particular airplane after a captain has refused to use that plane because he was not satisfied it was sufficiently safe.

"They ask another captain, and finally they'll find someone who says, 'Sure, I'll fly it.' I'm not saying the one who agrees to fly that plane is foolish; he may just think it's not a serious problem, or his skills are such that he can do it. But I don't want any part of it after I've heard a captain turn a plane down. I'll stick with the chicken every time."

This whole area is one that needs careful study by an outside agency.

And while they're looking at it, they ought to include the "I've got to get home" syndrome. Most often this is applied to private pilots, but anyone who has flown as a crew member for several years can tell you that senior pilots are also susceptible to the problem. I don't want to overstress this problem, however, because—as I said several times before

LOVE OF FLYING

—you have to remember that no pilot ever forgets that in a crash he is likely to be the first to "buy the farm."

Another area of interest should be the training of the crew in regard to safety measures in the event of accident. I've not spent time on this because it is not an area where I have much firsthand knowledge. The airlines will tell you that an average crew is trained to evacuate a normal-sized jet in two minutes or less. And Robert J. Serling, a prolific aviation writer with numerous books and novels to his credit, once took part in just such a demonstration and came away convinced. In a recent landing accident along the lower Atlantic seaboard, the crew of an Eastern Airlines jet, led by an experienced stewardess who prides herself on her knowledge of evacuation technique, got a planeload of people onto the ground in forty-five seconds! Still, recent reports have questioned the authenticity of the equipment used to simulate these techniques, claiming that the door handle used in the mock-up could be opened with only normal force, whereas a *real* door required one hell of a lot of muscle. Again, it's an area that needs looking into.

Finally, in regard to pilots, there is one recommendation I would like to stress. It has to do with the fact that in every commercial airliner cockpit, there is still just one boss—the captain. I know that many pilots, especially senior captains, disagree with me, but I feel that there should be two four-stripers in every plane. Actually, the hard economic times have made this happen, and in some cases with certain large airlines even the engineer is or was a captain. But I'm talking about the situation where you have two men up front who both have gone as far as they can go up the company ladder, so that one need not feel he has to defer to the judgment of the other for fear of jeopardizing his eventual promotion.

I have thought many times that perhaps 514 could have been saved if it had not been the captain who insisted that "cleared for the approach" meant what *he* thought it meant, and not what the dumb sheet seemed to indicate. If the copilot had made that error, would there have been any hesitancy on the part of any captain to set him straight? I doubt it, at least not if the captains I've met are any indication.

But I do not want to demean, by implication, the ability or the memory of the men in the front of that ill-fated plane. No one knows all of what happened up there in the midst of the horrendous difficulties, and we have only their recorded words, not their thoughts.

Yet I want to make the point that *captaincy* is very definitely a state of mind. The airlines will tell you that by having two experienced pilots in the cockpit they have accomplished full human redundancy. That's hogwash. A copilot (even one who has earned the fourth stripe) is not a captain. His state of mind is not captaincy. If the copilot decides to take the plane away from his captain, he had better have a very, very good reason.

Less deference and more democracy could provide an added safety buffer. I'm not suggesting anything radical—one man will always have to be in control of the plane and the other assisting him. But the assistant should never be afraid to overrule his boss, for his own life and that of many other people may ride on that decision.

The Polysemantics Problem

In the summer of 1976, in July to be exact, the NTSB made a public recommendation that was both simple and frightening. The board recommended that a dictionary of terms be drawn up so that pilots and controllers would be able to speak the same language!

I use the word "simple" because this good idea is so terribly obvious; and I use the word "frightening" because apparently no one thought it was necessary until 514 went into Mount Weather.

I recommend—no, I *urge*—that all the interested parties meet as soon as possible and begin to draw up such a dictionary if they agree one is necessary. And apparently one is.

The (Improved) World of General Aviation

The most pressing need in this area is to ensure a standard of excellence for all flight schools across the country—and in Alaska and Hawaii. Because flight is regulated (and regulated and regulated and regulated) by the government, there is no excuse for allowing substandard flight schools to exist. It should be clear by this time that flight school operators who stand to gain by getting a student through his or her training in record time must bear the burden of proof that they are exercising the proper standard of care and caution. I do not see government control in this area as an example of rampant Big Brotherism. A "pilot" who is graduated too soon is either imbued with false confidence or is incredibly stupid (because he or she *knows* it was all a bit too quick for true mastery). And both of these states all but guarantee that the pilot will very soon wander into danger.

Not only should the VFR training be standardized and brought up to the level of the best schools, but the amount of "indoctrinary" instrument training should be increased. It is just not enough to equip people to fly on clear days —anyone with minimum skills can learn to do that. They should all be given enough training to know how to get out of inadvertent trouble. A private pilot who panics when suddenly enveloped by a cloud bank is not a properly trained pilot.

The FAA has generally earned good marks in this area —with the exception of having looked the other way while certain flight schools continue to operate—and all that is really needed is for the agency to tighten up.

Doing Something About the Weather

As it presently stands, and especially in regard to the major airline flights, the pilot has almost the final word as to the seriousness of the weather conditions. By that I mean he can stretch himself and his airplane to the limits of their respective abilities in order to get one good shot at the rabbit running. As a pilot, I agree with that point of view. But as a passenger, there have been times when, in retrospect, I've wished the airport management would have closed down the field and saved a planeful of people a gutful of worry and anxiety.

It is one thing for a private pilot (with an instrument rating, of course) flying a ship filled with all the latest electronic goodies and one passenger to shoot at least one approach. And, given the skill of the average airline pilot, the situation is in essence not much different for a Rockwell Commander than for a Boeing 727. But the paying public deserves a higher standard of care. And I think that in the long run most passengers would not mind terribly if they had to end up in Milwaukee rather than Chicago, Boston rather than Hartford, or even Miami rather than Atlanta. But the airlines have built up a myth of perfection that could, in itself, become dangerous.

It is not without good reason that the FAA demands all pilots file an alternate flight plan. There are days, and nights, when you just can't land where you would like to.

Air Crash Litigation

I have one main recommendation in this area. It involves the problem of multi-district litigation—the problem of conflicting state laws governing suits arising from the same crash. I would like to see a blue ribbon committee of the American Bar Association, or a well-known legal think tank like the American Law Institute, take up this question and study it thoroughly. I would expect that if they did, they would decide that this is one area where we definitely need one set of laws for the entire country. "Point of origin" shouldn't govern the amount of damages.

Who Speaks for the People?

As I said in Chapter 4, I have never been much for collective action, but I learned from working with the controllers that there are times when semi-rugged individualism doesn't even begin to make it. The other side of the coin, however, is that there are times when unionizing is too difficult (or too expensive) to have any real effect.

I've watched for several years to see if Ralph Nader's aviation arm—the Aviation Consumer Action Project, or ACAP—would have much clout. Unfortunately, its influence has been limited, in direct proportion I would suppose, to its lack of money, manpower, and experience. But it was and is a step in the right direction, if for no other reason than that it is so diametrically opposed to the wealth and power of the airline industry.

I recommend the formation of a special office, within or related to the FAA, that would function as an ombudsman.

A wild idea? I think not. What else would the Consumer Protection Agency have been if that proposed legislation had passed but an ombudsman on a grander and more organized

LOVE OF FLYING [211]

scale. It would have been a voice *for* the people within the government itself. And that's exactly what I feel is needed in the area of aviation safety.

Clearly, the Congress is going to have a hard time regulating the FAA; and just as clearly the FAA is not about to police itself. So why not place a person—a man like Chuck Miller, who used to run the NTSB until he spoke out too often and too clearly, comes immediately to mind—in a position of authority as a vocal house critic. And give him the power to follow up the complaints he receives or initiates. Give such a person the power to "go between" the people and the government. I would be willing to bet that if we were lucky enough to find the right person, we would soon get results.

What could we lose by trying?

* * *

Ann Darr's poem, "Cleared for Approach, Cleared for Landing," is divided into ten parts. One line in the ninth stanza reads, "We know and we do not/want to know what we know. Everyone/has second sight. . . ."

The poem concludes with this brief section:

> And then it was I heard the horn
> blowing through the fog, and I did not cry
> out, but gave all power and rose up
> and cleared the mountain and the weather
>
> and knew all that was left to me.

* * *

Let us all hope we can make that same testament of faith. If we can't, we had best do something about it, and soon.

Epilogue

On March 10, 1977, the worst crash in the history of aviation took place. Five hundred and eighty-two people lost their lives, either immediately or within days of the crash, which, ironically, took place not in the air but on the ground.

Until that day, the name Tenerife meant a beautiful island, the largest of the seven Canary Islands off the coast of Northwest Africa. Today, Tenerife is synonymous with tragedy, for it was at the smaller of Tenerife's two airports that the crash occurred. Another of the ironies with which this accident is dotted is that the island's main airport (Las Palmas) was closed because of a terrorist bomb scare; all flights, even those of the big jumbo jets, had been diverted to the smaller Los Rodeos, where fog and crowded runways were the backdrop for disaster.

It was early afternoon—cool, windy, and so foggy that only a sixth of the two-mile-long runway was visible. Neither the cockpit crew of planes waiting to take off, nor the controllers in the tower, could see anything but dense fog. There was no ground radar; all traffic on the ground was being moved by radio communication with the tower. The number of diverted flights had flooded the small airport, and the lone taxiway was blocked, which meant the single runway had to be used for both takeoffs and taxiing.

Two Boeing 747s scheduled for Las Palmas had landed at Los Rodeos, refueled, and prepared to take off again. The 232-foot-long KLM 4805 was the first to taxi out. It moved slowly down the runway, and swung around into takeoff position.

The tower then cleared Pan Am to taxi to the opposite end of the runway. When it had done so, Pan Am was instructed to move up the runway—toward the KLM plane, which was waiting for take-off clearance—but to turn off at "the third intersection." There were four "intersections," or ramps (C1-C4) connecting the runway to the taxi strip.

As Pan Am rolled through the fog toward KLM's end of the runway, KLM checked with the tower to make certain that Pan Am would be moving off at the third ramp.

"Affirmative. One, two, three. The third one, sir," said the controller in the tower.

Next, KLM asked for clearance of its flight path, but *not* for takeoff clearance. The information crackled through the speakers, and, as the regulations require, KLM repeated it.

Then, only moments later, the tower told KLM to stand by for takeoff clearance, adding, "I will call you back."

Right after that the tower told Pan Am to report when it was clear of the runway. Pan Am replied that it would.

As these last exchanges were concluding, the big Pan Am Boeing was rolling *past* ramp C-3 and heading for C-4. (Later, it would come out that Pan Am had not counted C-1, because it was blocked, as an open "intersection," and thus was heading for the third *available* intersection.)

Everything was still shrouded in fog. Because of a malfunction, not even the strip of white lights in the middle of the concrete runway was lit. Pan Am 1736 was about 450 feet away from C-4 when its crew saw lights shining through the fog. Then the lights became progressively bigger—and the crew realized its peril. The KLM 747 was roaring down the runway, straight at the Pan Am jet.

The Pan Am captain tried to drive his huge plane off into the grass between C-3 and C-4, at the same time screaming into his mike, "We are still on the runway. What's he doing? He'll kill us all."

When the crew of the KLM spotted the Pan Am jet dead in its path, the Dutch captain made an instantaneous (and heroic) move. By pulling up the nose, he tried to "jump" his plane over the other 747. He came within 25 feet of making it.

We all remember the color pictures of the crash that appeared in newspapers and magazines, so I will not dwell on the details of what happened after the two giant airships collided. No one on the KLM 747 survived. Sixty-seven people from the Pan Am flight were saved. The death toll was the highest in the history of aviation.

What happened? What went wrong? Man or machine? Apparently, this time it was the fault of man.

One more irony: when KLM learned of the accident, the

EPILOGUE [215]

company tried unsuccessfully to reach one of its most able and distinguished pilots to investigate the crash. Captain Jacob Veldhuizen Van Zanten, fifty-one, had been with KLM for twenty-five years. Not only had he trained other KLM pilots, but he had also appeared in the company's commercial advertising as the epitome of skill and experience. The reason KLM couldn't reach Captain Veldhuizen Van Zanten was that he had been the pilot who had driven KLM 4805 into Pan Am 1736.

Spain, which controls the Canaries, has still to release its long-overdue report of the crash. More significantly, the Spaniards have not released the tape of the KLM cockpit voice recorder or the tape that includes the tower's communication with KLM. (However, Pan Am's tape is available, and it contains what the tower said to KLM because both planes were on the same frequency.) What was said inside the KLM cockpit, by one crew member to another, may be highly significant.

Initially, there was some suspicion that the Spanish controllers might have been at fault, because of their lack of experience with airliners in general and jumbo-jets in particular. Perhaps, some said, they had confused the clearances and thereby caused the accident. But the Pan Am tape indicates that their transmissions were in perfectly adequate English. It is clear that Pan Am understood the words spoken (if not the meaning of the "third intersection"), and the consensus of opinion among captains I've talked to is that they would have understood that they had no clearance to go—there was still another plane on the runway.

Certain facts are clear. Pan Am was cleared to be on the runway by the tower; the tower knew Pan Am was on the runway; and the tower's transmissions to Pan Am were heard in the cockpit of the KLM jet. Still unclear is the tragedy's central issue: Why did the highly experienced Veldhuizen Van Zanten begin to roll toward takeoff without being cleared by the tower (and even the Dutch agree that this is precisely what he did)? As yet, no one knows.

Unfortunately, this kind of accident is not entirely without precedent. I can think of two somewhat similar accidents, one in Boston in 1960 and another in Chicago seven years later. What makes them similar is that in all cases communications were at fault.

In the first accident, a freakish and almost unique jumbling of radio messages caused a captain to believe, mistakenly, that he had been cleared for takeoff, and his jet was struck by another airliner that was landing (with proper clearance). In Chicago, the tower told the pilot of a just-arrived flight to taxi to the "3-2" holding area.

He began to do so, not realizing that he was taxiing across the active runway, and a DC-9 in the process of taking off crashed into him. The problem: there were two "3-2" holding areas, one to the left of runway 3-2, and one to its right. The controller had one in mind, the captain had the other.

The crashes in Boston and Chicago were precedents for Tenerife in the sense that all three indicate lack of clear, systematic follow-up.

As I've said earlier: redundancy is the only thing that saves the aviation business. Having two of so many expensive pieces of machinery on a big airplane may seem needlessly expensive, but comes the time one fails and the other clicks in to save you and everyone else on board, you become eternally grateful. We haven't yet found a way to make the wing redundant, so that a plane can lose one and still fly, any more than a helicopter can stay in the air after losing a rotor blade, but almost everything else on an airplane is redundant.

After the 1960 crash at Logan Airport, the NTSB made several recommendations, but those I remember most vividly called for more redundancy and better communications. Because of that accident, the one in Chicago, and now the holocaust known as Tenerife, I must reiterate the importance of that advice. I believe the following recommendation to be vitally important:

> In conditions of low visibility, all transmissions should be made by the tower *with a readback* from the captain (meaning, he should repeat verbatim what the tower has just told him), and then the same message should be sent to the copilot or first officer, also followed by a readback.

This would ensure that both pilot and co-pilot heard the same thing and both acknowledged it in the same words. And the terms used should be highly precise and should be the same worldwide, with no variations.

If this kind of constructive redundancy in communications was to be adopted, the likelihood of a repetition of what happened in Tenerife, in Chicago, and (though to a lesser degree) in Boston would be greatly decreased. Communications is the thread that ties all three of these accidents together.

It now looks as if, when all the facts are in, the responsibility for the tragedy of Tenerife will be placed on the grave of a man who was by all accounts a superior pilot. I say that reluctantly, but I am compelled to say it because of the rumor I first heard several months after the accident, and later had verified by a senior United States aviation official.

It seems that according to the KLM cockpit tape, as the KLM 747 began to roar down the runway, oblivious of the Pan Am jet in its foggy path, the First Officer said to Captain Veldhuizen Van Zanten in Dutch, "Captain, are you sure we should be rolling?"

Some pages back, I made a recommendation that every cockpit of every commercial airliner should carry two captains. Of all the recommendations in this book, that one drew the sole adverse criticism from airline captains. Such a practice, they claimed, would surely undermine traditional "left-seat discipline."

I have a healthy respect for discipline, but it can be a curse as well as a benefit—as in the case of Tenerife.

The co-pilot of KLM 4805 could only question his captain as a subordinate. What if there had been another *captain* in the right seat? Would he have been so deferential? I doubt it.

If he was a seasoned captain, he would have—as a natural response to the situation—grabbed the throttles and said, "Sorry, Captain, but *I'm* not convinced that the runway is clear."

Had that bit of revised history taken place, had another senior captain been in the position of challenging the most prestigious pilot of one of the most prestigious airlines in the world, then there would have been no accident. And two great big airplanes would have been saved—plus 582 lives.

F. Lee Bailey, one of the most famous attorneys in American history and author of such bestselling books as FOR THE DEFENSE and THE DEFENSE NEVER RESTS became a pilot at the age of twenty, thanks to the U.S. Navy and, later, the Marines. He has flown his own planes ever since, from Cessnas to Lear Jets and helicopters. In addition to his better-known legal activities, he currently operates an air charter service and owns a company that makes helicopters.

John Greenya is a well-known professional writer. He was co-author of FOR THE DEFENSE with F. Lee Bailey and of MAXINE CHESHIRE, REPORTER with Maxine Cheshire of the *Washington Post*.

Big Bestsellers from SIGNET

☐ **KRAMER VERSUS KRAMER** by Avery Corman.
(#E8282—$2.50)

☐ **EVEN BIG GUYS CRY** by Alex Karras.
(#E8283—$2.25)*

☐ **VISION OF THE EAGLE** by Kay McDonald.
(#J8284—$1.95)*

☐ **HOMICIDE ZONE FOUR** by Nick Christian.
(#J8285—$1.95)*

☐ **CRESSIDA** by Clare Darcy. (#E8287—$1.75)*

☐ **BELIEVING IN GIANTS** by Claire Vincent.
(#E8289—$2.25)*

☐ **THE NOBLE PIRATE—The Highwayman #3** by Raymond Foxall. (#E8291—$1.75)*

☐ **THE INFORMANT** by Marc Olden. (#E8393—$1.75)*

☐ **DANIEL MARTIN** by John Fowles. (#E8249—$2.95)

☐ **THE EBONY TOWER** by John Fowles. (#E8254—$2.50)

☐ **THE FRENCH LIEUTENANT'S WOMAN** by John Fowles.
(#E8066—$2.25)

☐ **MISTRESS OF OAKHURST—Book II** by Walter Reed Johnson. (#J8253—$1.95)

☐ **OAKHURST—Book I** by Walter Reed Johnson.
(#J7874—$1.95)

☐ **RIDE THE BLUE RIBAND** by Rosalind Laker.
(#J8252—$1.95)

☐ **THE SILVER FALCON** by Evelyn Anthony.
(#E8211—$2.25)

☐ **I, JUDAS** by Taylor Caldwell and Jess Stearn.
(#E8212—$2.50)*

☐ **THE RAGING WINDS OF HEAVEN** by June Shiplett.
(#J8213—$1.95)*

☐ **THE TODAY SHOW** by Robert Metz. (#E8214—$2.25)

☐ **HEAT** by Arthur Herzog. (#J8115—$1.95)*

☐ **THE SWARM** by Arthur Herzog. (#E8079—$2.25)

* Price slightly higher in Canada

More Big Bestsellers from SIGNET

- ☐ **THE SERGEANT MAJOR'S DAUGHTER** by Sheila Walsh.
  (#E8220—$1.75)
- ☐ **BLOCKBUSTER** by Stephen Barlay. (#E8111—$2.25)*
- ☐ **BALLET!** by Tom Murphy. (#E8112—$2.25)*
- ☐ **THE LADY SERENA** by Jeanne Duval.
  (#E8163—$2.25)*
- ☐ **LOVING STRANGERS** by Jack Mayfield.
  (#J8216—$1.95)*
- ☐ **BORN TO WIN** by Muriel James and Dorothy Jongeward.
  (#E8169—$2.50)*
- ☐ **BORROWED PLUMES** by Roseleen Milne.
  (#E8113—$1.75)
- ☐ **ROGUE'S MISTRESS** by Constance Gluyas.
  (#E8339—$2.25)
- ☐ **SAVAGE EDEN** by Constance Gluyas. (#E8338—$2.25)
- ☐ **WOMAN OF FURY** by Constance Gluyas.
  (#E8075—$2.25)*
- ☐ **BEYOND THE MALE MYTH** by Anthony Pietropinto, M.D., and Jacqueline Simenauer. (#E8076—$2.50)
- ☐ **CRAZY LOVE: An Autobiographical Account of Marriage and Madness** by Phyllis Naylor. (#J8077—$1.95)
- ☐ **THE PSYCHOPATHIC GOD—ADOLF HITLER** by Robert G. L. Waite. (#E8078—$2.95)
- ☐ **THE SERIAL** by Cyra McFadden. (#J8080—$1.95)
- ☐ **MARATHON** by Jules Witcover. (#E8034—$2.95)
- ☐ **THE RULING PASSION** by Shaun Herron.
  (#E8042—$2.25)
- ☐ **CONSTANTINE CAY** by Catherine Dillon.
  (#J8307—$1.95)
- ☐ **WHITE FIRES BURNING** by Catherine Dillon.
  (#J8281—$1.95)
- ☐ **THE WHITE KHAN** by Catherine Dillon.
  (#J8043—$1.95)*
- ☐ **THE MASTERS WAY TO BEAUTY** by George Masters with Norma Lee Browning. (#E8044—$2.25)

*Price slightly higher in Canada

Have You Read These SIGNET Bestsellers?

- ☐ **KID ANDREW CODY AND JULIE SPARROW** by Tony Curtis. (#E8010—$2.25)*
- ☐ **WINTER FIRE** by Susannah Leigh. (#E8011—$2.25)*
- ☐ **THE MESSENGER** by Mona Williams. (#J8012—$1.95)
- ☐ **FEAR OF FLYING** by Erica Jong. (#E7970—$2.25)
- ☐ **HOW TO SAVE YOUR OWN LIFE** by Erica Jong. (#E7959—$2.50)*
- ☐ **HARVEST OF DESIRE** by Rochelle Larkin. (#E8183—$2.25)
- ☐ **MISTRESS OF DESIRE** by Rochelle Larkin. (#E7964—$2.25)*
- ☐ **THE QUEEN AND THE GYPSY** by Constance Heaven. (#J7965—$1.95)
- ☐ **TORCH SONG** by Anne Roiphe. (#J7901—$1.95)
- ☐ **ISLAND OF THE WINDS** by Athena Dallas-Damis. (#J7905—$1.95)
- ☐ **FRENCH KISS** by Mark Logan. (#J7876—$1.95)
- ☐ **CARIBEE** by Christopher Nicole. (#J7945—$1.95)
- ☐ **THE DEVIL'S OWN** by Christopher Nicole. (#J7256—$1.95)
- ☐ **MISTRESS OF DARKNESS** by Christopher Nicole. (#J7782—$1.95)
- ☐ **CALDO LARGO** by Earl Thompson. (#E7737—$2.25)
- ☐ **A GARDEN OF SAND** by Earl Thompson. (#E8039—$2.50)
- ☐ **TATTOO** by Earl Thompson. (#E8038—$2.50)
- ☐ **DESIRES OF THY HEART** by Joan Carroll Cruz. (#J7738—$1.95)
- ☐ **THE SHINING** by Stephen King. (#E7872—$2.50)
- ☐ **COMA** by Robin Cook. (#E8202—$2.50)

*Price slightly higher in Canada

NAL/ABRAMS' BOOKS ON ART, CRAFTS AND SPORTS

in beautiful, large format, special concise editions—lavishly illustrated with many full-color plates.

- [ ] **THE ART OF WALT DISNEY: From Mickey Mouse to the Magic Kingdoms** by Christopher Finch. (#G9982—$7.95)

- [ ] **DISNEY'S AMERICA ON PARADE: A History of the U.S.A. in a Dazzling, Fun-Filled Pageant,** text by David Jacobs. (#G9974—$7.95)

- [ ] **FREDERIC REMINGTON** by Peter Hassrick. (#G9980—$6.95)

- [ ] **GRANDMA MOSES** by Otto Kallir. (#G9981—$6.95)

- [ ] **THE POSTER IN HISTORY** by Max Gallo. (#G9976—$7.95)

- [ ] **THE SCIENCE FICTION BOOK: An Illustrated History** by Franz Rottensteiner. (#G9978—$6.95)

- [ ] **NORMAN ROCKWELL: A Sixty Year Retrospective** by Thomas S. Buechner. (#G9969—$7.95)

- [ ] **THE PRO FOOTBALL EXPERIENCE** edited by David Boss, with an Introduction by Roger Kahn. (#G9984—$6.95)

- [ ] **THE DOLL** text by Carl Fox, photographs by H. Landshoff. (#G9987—$5.95)

- [ ] **DALI . . . DALI . . . DALI . . .** edited and arranged by Max Gérard. (#G9983—$6.95)

- [ ] **THOMAS HART BENTON** by Matthew Baigell. (#G9979—$6.95)

- [ ] **THE WORLD OF M. C. ESCHER** by M. C. Escher and J. L. Locher. (#G9970—$7.95)

THE NEW AMERICAN LIBRARY, INC.,
P.O. Box 999, Bergenfield, New Jersey 07621

Please send me the SIGNET, MENTOR and ABRAMS BOOKS I have checked above. I am enclosing $_____ (please add 50¢ to this order to cover postage and handling). Send check or money order—no cash or C.O.D.'s. Prices and numbers are subject to change without notice.

Name_____

Address_____

City_____ State_____ Zip Code_____

Allow at least 4 weeks for delivery
This offer is subject to withdrawal without notice.

"WE ONLY HAVE ONE TEXAS"

People ask if there is really an energy crisis. Look at it this way. World oil consumption is 60 million barrels per day and is growing 5 percent each year. This means the world must find three million barrels of new oil production each day. Three million barrels per day is the amount of oil produced in Texas as its peak was 5 years ago. The problem is that it is not going to be easy to find a Texas-sized new oil supply every year, year after year. In just a few years, it may be impossible to balance demand and supply of oil unless we start conserving oil today. So next time someone asks: "is there really an energy crisis?" Tell them: "yes, we only have one Texas."

ENERGY CONSERVATION -
IT'S YOUR CHANCE TO SAVE, AMERICA

Department of Energy, Washington, D.C.

A PUBLIC SERVICE MESSAGE FROM NEW AMERICAN LIBRARY, INC.